# Introduction to Excel

## Fifth Edition

**DAVID C. KUNCICKY**
Bioreason, Inc., Santa Fe, NM

**RONALD W. LARSEN**
Montana State University, Bozeman, MT

PEARSON

Boston • Columbus • Indianapolis • New York
San Francisco • Upper Saddle River • Amsterdam
Cape Town • Dubai • London • Madrid • Milan
Munich • Paris • Montreal • Toronto • Delhi
Mexico City • São Paulo • Sydney • Hong Kong
Seoul • Singapore • Taipei • Tokyo

Vice President and Editorial Director, ECS: *Marcia J. Horton*
Executive Editor: *Holly Stark*
Editorial Assistant: *Carlin Heinle*
Executive Marketing Manager: *Tim Galligan*
Marketing Assistant: *Jon Bryant*
Permissions Project Manager: *Karen Sanatar*
Senior Managing Editor: *Scott Disanno*
Production Project Manager/Editorial Production Manager: *Greg Dulles*
Project Manager: *Pat Brown*
Creative Director: *Jayne Conte*
Designer: *Bruce Kenselaar*
Composition: *Jouve India*
Full-Service Project Management: *Pavithra Jayapaul, Jouve India*
Printer/Binder: *Edwards Brothers Malloy*
Cover Printer: *Edwards Brothers Malloy*
Typeface: *10/12 ITC New Baskerville Std-Roman*

© 2013, 2010, 2006, 2000 Pearson Education, Inc.
Upper Saddle River, New Jersey 07458

The author and publisher of this book have used their best efforts in preparing this book. These efforts include the development, research, and testing of the theories and programs to determine their effectiveness. The author and publisher make no warranty of any kind, expressed or implied, with regard to these programs or the documentation contained in this book. The author and publisher shall not be liable in any event for incidental or consequential damages in connection with, or arising out of, the furnishing, performance, or use of these programs.

Pearson Prentice Hall™ is a trademark of Pearson Education, Inc.

Printed in the United States of America.

**10 9 8 7 6 5 4 3 2**

Pearson Education Ltd., *London*
Pearson Education Australia Pty. Ltd., *Sydney*
Pearson Education Singapore, Pte. Ltd.
Pearson Education North Asia Ltd., *Hong Kong*
Pearson Education Canada, Inc., *Toronto*
Pearson Educación de Mexico, S.A. de C.V.
Pearson Education—Japan, *Tokyo*
Pearson Education Malaysia, Pte. Ltd.
Pearson Education, *Upper Saddle River, New Jersey*

CIP data is on file and available upon request.

ISBN 10: 0-13-308363-2
ISBN 13: 978-0-13-308363-7

# Contents

## 3 • FORMULAS AND FUNCTIONS 99

## 4 • WORKING WITH CHARTS 149

# About This Book

The *Introduction to Excel* text is designed for first- or second-year engineering majors in computer applications courses. Excel® provides a rich environment for many engineering calculations, and *Introduction to Excel* is designed to get engineering students up and running with Excel as quickly as possible. With the fifth edition:

- The text has been updated for Excel 2010.
- Command sequences for Excel 2010, 2007, and 2003 are provided since all three editions of Excel are still in use.
- All images have been updated to illustrate Excel 2010 running in Windows 7.
- Annotations on screen captures have been improved.
- The new approach to print and print preview used in in Excel 2010 is illustrated.
- The use of the significantly improved Solver that is now included with Excel 2010 is covered.

# 1

# Microsoft Excel Basics

## Sections

## Objectives

*After reading this chapter, you should be able to perform the following tasks:*

- Describe how spreadsheets are used by engineers.
- Identify the main components on the Excel screen.
- Name at least two ways to access help for Excel.
- Create and save a new worksheet.
- Open and edit an existing worksheet.
- Undo mistakes.
- Perform spelling and grammar checks on text items.
- Preview and print a worksheet.

## 1.1 INTRODUCTION TO WORKSHEETS

A *spreadsheet* is a rectangular grid composed of addressable units called *cells*. A cell is addressed by referencing its column letter and row number. A cell may contain numerical data, textual data, or *formulas (equations)*.

Spreadsheet application programs were originally intended to be used for financial calculations. The original electronic spreadsheets resembled the paper spreadsheets of an accountant. One characteristic of electronic spreadsheets that gives them an advantage over their paper counterparts is their ability to automatically recalculate all dependent values whenever a parameter is changed.

Over time, more and more functionality has been added to spreadsheet programs, like *Excel*. A variety of mathematical and engineering functions now exist within Excel. Numerous analytical tools are also available, including scientific and engineering tools, statistical tools, data-mapping tools, and financial-analysis tools. Auxiliary functions include a graphing capability, database functions, and the ability to access the World Wide Web.

As an engineering student, you may find that an advanced spreadsheet program such as Microsoft Excel will suffice for many of your computational and data presentation needs. You will still need a word processor, such as Microsoft Word, for working with reports and other documents, but tables and charts may be easily exported from Excel into Word.

Excel also has some capability for database management. However, if you wish to manage large or sophisticated databases, a specialized database application such as Microsoft Access or MySQL is preferable.

In addition, Excel has fairly sophisticated mechanisms for performing mathematical and scientific analyses. For example, you can use the Analysis ToolPak in Excel to perform mathematical analysis. If the analysis is large or very sophisticated, however, you may want to use a specialized mathematical or matrix package such as Mathcad® or Matlab®.

The same principles hold for graphing (SigmaPlot®, Origin®) or statistical analysis (SAS®, SPSS®). Excel is a general tool that performs many functions for small- to medium-sized problems. As the size or sophistication of the function increases, other tools may be more applicable.

Microsoft Excel uses the term *worksheet* to denote a spreadsheet. A worksheet can contain more types of items than a traditional paper spreadsheet. These include charts, links to web pages, Visual Basic programs, and macros. We will treat the terms *worksheet* and *spreadsheet* synonymously in this text. Worksheets stored together in a file are called a *workbook*.

## 1.2 HOW TO USE THIS BOOK

This book is intended to get you, the engineering student, up and running with Excel 2010 as quickly as possible. (References to Excel 2007 and 2003 are provided as well, since many still use these versions.) Examples are geared toward engineering and mathematical problems. Try to read the book while sitting in front of a computer. Learn to use Excel by re-creating each example in the text. Perform the instructions in the boxes labeled **PRACTICE**.

The book is not intended to be a complete reference manual for Excel. It is much too short for that purpose. Many books on the market are more appropriate for use as complete reference manuals. However, if you are sitting at the computer, one of the best reference manuals is at your fingertips. The online Excel help tools provide an excellent resource if properly used. These help tools are described later in this text.

## 1.3 TYPOGRAPHIC CONVENTIONS USED IN THIS BOOK

Throughout the text, the following conventions will be used:

### Selection with the Mouse

The book frequently asks you to move the mouse cursor over a particular item and then click and release the left mouse button. This action is repeated so many times in the text that it will be abbreviated as follows:

Choose **Item**.

If the mouse button is not to be released or if the right mouse button is to be used, then this will be stated explicitly.

A button, icon, or menu option that you are to select with the mouse will be printed in boldface font. A key you should press will also be printed in boldface font. For example, if you are asked to choose an item from the menu-strip (called the *Ribbon*) at the top of the screen labeled *Paste,* then it will be written as follows:

Choose **Paste** from the Ribbon's *Home tab*.

### Multiple Selections

The book frequently refers to selections that require more than one step. For example, to see a print preview, perform the following steps:

1. Choose **Cell Styles** from the Excel Ribbon.
2. Choose **Normal** style from the drop-down menu.

Multiple selections will be abbreviated by separating choices with a right arrow. For example, the two steps listed will be abbreviated as follows:

Choose **Cell Styles → Normal** from the Ribbon.

### Multiple Keystrokes

If you are asked to simultaneously press multiple keys, the key names will be printed in bold italic font and will be separated with a plus sign. For example, to undo a typing change, you can simultaneously press the **Ctrl** key and the **Z** key. This will be abbreviated as follows:

To undo typing, press **Ctrl + Z**.

### Key Terms

The first time a key term is used, it will be italicized. Key terms are summarized at the end of each chapter.

### Literal Expressions

A word or phrase that is a literal transcription will be printed in bold. For example, the Title bar at the top of the screen should contain the text **Microsoft Excel**. Another example is the literal name of a box or menu item, such as the following:

Check the box labeled **Equal To**.

## 1.4 UNDERSTANDING THE EXCEL 2010 SCREEN

This section introduces you to the Microsoft Excel screen. To start the Excel program, use the Windows Start menu (illustrated in Figure 1.1):

Start → Microsoft Office → Microsoft Excel 2010

**Figure 1.1**
Launching Excel from the Start menu.

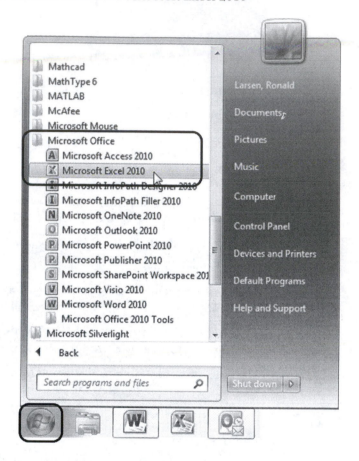

A screen that resembles Figure 1.2 will appear.

**Figure 1.2**
The Excel screen, part 1.

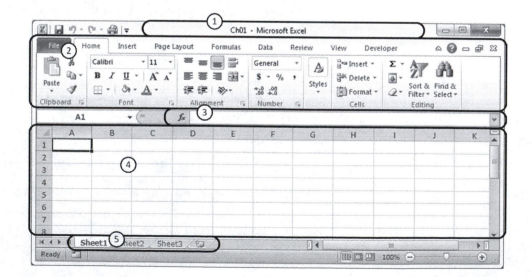

We'll now discuss each of the components on the screen. The Excel screen consists of a number of components, including the following:

1. Title bar
2. Ribbon (Menu bar in Excel 2003)
3. Formula bar
4. Work area
5. Sheet tabs

In Figure 1.3, additional screen features are identified:

1. Quick Access Toolbar
2. Excel window control buttons
3. Worksheet control buttons
4. Excel Help button
5. Ribbon toggle (displays and hides the Ribbon)
6. Name box
7. Status bar
8. Zoom control
9. Excel window resize handle

**Figure 1.3**
The Excel screen, part 2.

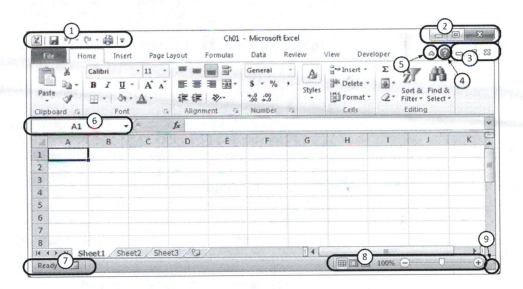

Try to become familiar with the names of these components as we proceed, as we will use these names throughout the book. Working generally from top to bottom, the major components will be discussed in the following sections.

### 1.4.1 Title Bar

The bar at the top of the screen is called the *Title bar*. The Title bar contains the name of the worksheet currently being edited, Ch01.xlsx in Figure 1.2. If you are working in an unsaved workbook, the default name **Book1** will appear in the Title bar.

The Title bar contains a number of useful buttons and features (from left to right):

- *Quick Access Toolbar*—The small collection of buttons just to the right of the Office button is the *Quick Access Toolbar*. This area is designed for your use, to add buttons for the features that you use most often. The small down arrow to the right of the Quick Access Toolbar opens a menu that you can use to customize the toolbar.

- *File Name*—The name of the workbook that is being edited is displayed in the center of the Title bar when the workbook has been maximized to fill the entire work area. If the current workbook is not maximized, then it will be displayed in its own window in the work area, with the file name shown at the left side of the workbook window's Title bar, as illustrated in Figure 1.4.

**Figure 1.4**
Multiple workbooks can be opened in the work area.

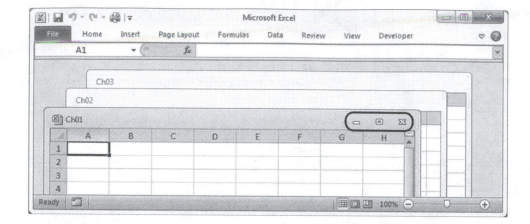

- *Control Buttons*—The three buttons at the right side of the Title bar are called the *Control buttons*. They are used to control the way the Excel window is displayed.
    - *Minimize Button*—The small flat line is the **Minimize** button. If you click the *minimize button* the Excel window will disappear from your desktop, except for the Excel icon on the Taskbar, which usually appears at the bottom of the desktop. Click the Taskbar icon to restore the Excel window on your screen.
    - *Maximize/Restore Window Toggle Button*—The middle button is a toggle button that changes the display back and forth between two options.
        - If the window is not maximized, then the middle button maximizes the window (causes it to fill the entire desktop).
        - If the window is already maximized, then the middle button restores whatever size the window was before it was last maximized.
    - *Close Button*—The rightmost button on the Excel Title bar is the **Close** *button* (shaped like an x). Closing the Excel window is equivalent to exiting the Excel program. If you have made changes to the workbook you will be asked if you want to save the workbook before exiting.

When the worksheet is maximized to fill the entire Excel work area (as in Figure 1.3), the worksheet control buttons appear directly below the Excel window control buttons.

When the worksheets are not maximized, the control buttons for each worksheet are shown at the right side of the worksheet's Title bar as indicated in Figure 1.4.

### 1.4.2 Ribbon

The *Ribbon* was a new feature in Office 2007 and replaced the Menu bar, most toolbars, and some *dialog boxes*. The Ribbon attempts to get everything you need to use Excel right where you can get at it quickly. It is context sensitive, so that when you are editing a chart, for example, the *Ribbon tabs* related to working with charts are activated. The Ribbon can be minimized, as shown in Figure 1.4, but is more commonly used in the expanded form as shown in Figure 1.5.

**Figure 1.5**
The Ribbon's Home tab.

The Ribbon is made up of a number of *tabs*:

- *File tab*—used to open, save, and print workbooks, and to set Excel Options.
- *Home tab*—very commonly used commands for formatting and sorting.
- *Insert tab*—used to insert objects such as charts and hyperlinks.
- *Page Layout tab*—used to modify entire sheets (apply themes, set print area, etc.).
- *Formulas tab*—used to insert functions and manage defined names of cells and cell ranges.
- *Data tab*—provides access to sorting and filtering features, and data analysis tools (if activated).
- *Review tab*—used to add comments and track changes to a worksheet.
- *View tab*—used to change the display magnification (zoom), and to show or hide features such as the Formula bar and *gridlines*.
- *Developer tab*—not displayed by default, provides access to the Visual Basic editor and macros.
- *Add-Ins tab*—not displayed unless you have installed Excel Add-Ins. Excel Add-Ins are programs written for Excel by other software companies that are intended to extend the capabilities of Excel.

Most of the features you will need for day-to-day problem solving will be on the Home tab.

> **Note:** Microsoft Office 2007 applications, including Excel 2007, included something called the *Office button*, a big round button at the left end of the Title bar. It has been replaced by the File tab on the Ribbon in Excel 2010. Use the File tab (File menu in Excel 2003) to:
>
> - Open workbooks
> - Save workbooks
> - Print workbooks
> - Set Excel Options

Each tab is divided into *Groups* of related buttons, selection lists, and menus. For example, the Font group on the Home tab (shown in Figure 1.5) contains drop-down lists for font size and style, toggle buttons for font attributes (bold, italic), and combination buttons (buttons with a small down arrow on the right side) for setting background (fill) and font colors. Clicking the button applies the color shown on the button. Clicking the down arrow opens a color palette so that you can select a color.

When this text instructs you to use a Ribbon option, it will be in the following general form:

**Tab → Group → Drop-down Menu → Button**

### 1.4.3 Formula Bar

The *Formula bar*, located just below the Ribbon, displays the formula (or text, or value) in the currently selected cell (called the *active cell*). In Figure 1.6, cell B3 is the active cell, and contains the formula:

$$= 3 + 4$$

When cell B3 is selected, the result of the calculation is displayed in the cell (as shown in Figure 1.6) and the cell contents (the formula) is displayed in the Formula bar.

**Figure 1.6**

The Formula bar displays the contents of the active cell.

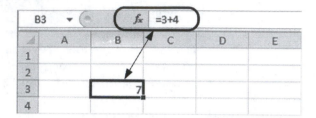

When you are entering a formula, you can type in the Formula bar or directly into the cell that will hold the formula. Most people enter formulas directly into the cells, but the Formula bar can be useful when entering a formula in a cell near the right edge of the work area.

The left side of the Formula bar is called the *Name Box*. The Name Box displays the name of the active cell. In Figure 1.6, the Name Box appears in the top-left corner and is displaying "B3" since that is the active cell.

The **Insert Function** *button* also resides on the Formula bar. The icon on the **Insert Function** button shows $f_x$ a common nomenclature for "function."

Click in cell C3 to make it the active cell, and then click on the **Insert Function** button. The Insert Function dialog box will appear, as shown in Figure 1.7. From the

**Figure 1.7**

The Insert Function dialog box.

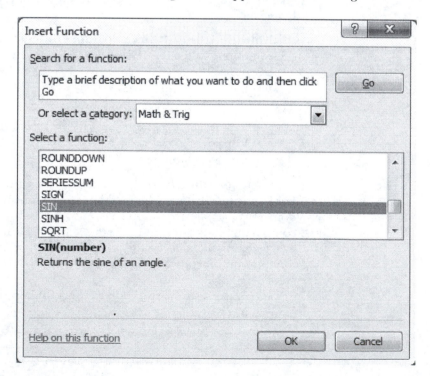

Insert Function dialog box, you can choose a function category and function name. In Figure 1.7, we have chosen the category **Math & Trig** and the function **SIN**.

Near the bottom of the Insert Function dialog box, a brief description of the function is displayed. The dialog box also has a *search* feature to help you locate a function. There are over 200 built-in functions available in Excel.

Choose the **SIN** (abbreviation for sine) function, then click **OK**. The Function Arguments dialog box will appear, as shown in Figure 1.8. This dialog prompts for the arguments to the named function.

**Figure 1.8**
The Function Arguments dialog box for the SIN function.

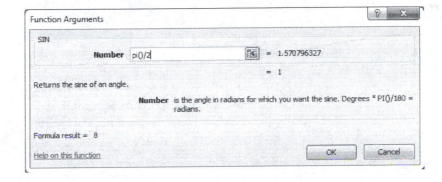

A short explanation about the expected arguments appears in the bottom of the window. In this case, the SIN function takes its arguments in radians. The formula for converting radians to degrees is also displayed.

The arguments may be a range of cells, numbers, or other functions. Type

$$pi()/2$$

as the **Number** argument. The effect of this is to call another built-in function, named *PI* (returns the value of $\pi$), and divide the result by 2.

When you click **OK**, the Function Arguments dialog box will disappear and the formula

$$= SIN(PI()/2)$$

will be entered into the active cell (cell C3). The result ($\sin(\pi/2) = 1$) is displayed in cell C3, as illustrated in Figure 1.9.

**Figure 1.9**
The formula = SIN(PI( )/2) entered in cell C3.

| C3 | ▼ | $f_x$ | =SIN(PI()/2) | | |
|---|---|---|---|---|---|
| | A | B | C | D | E |
| 1 | | | | | |
| 2 | | | | | |
| 3 | | | 7 | 1 | |
| 4 | | | | | |

### 1.4.4 Work Area

The *Work area* (also called the *Workbook window*) is the area on the screen where data are entered and displayed. The work area contains one or more worksheets.

The maximum size for a worksheet is 1,048,576 rows by 16,384 columns (Excel 2003: 65,536 × 256). The columns are labeled A, B, C, . . . , AA, AB, . . . , AAA, AAB, . . . , XFD, and the rows are labeled 1, 2, 3, . . . , 1048576.

A single cell can be selected by placing the mouse over the cell and clicking the mouse. The selected cell is called the *active cell*. A **range** of cells can be selected by holding the left mouse button down and dragging it over the selected cell range. When a cell range is selected, the first cell selected is the active cell. In Figure 1.10, the cell range B2:C4 is selected, and cell B2 is the active cell.

**Figure 1.10**

Selected cell range B2:C4, with active cell B2.

An entire column can be selected by clicking the left mouse button on the **column heading**. An entire row can be selected by clicking on the **row heading**. The entire worksheet can be selected by clicking on the heading in the top-left corner of the workbook.

### 1.4.5 Sheet Tabs

The *sheet tabs* are located at the bottom of the displayed worksheet, as shown in Figure 1.11. You can have more than one worksheet in a workbook. The sheet tabs identify all of the worksheets in current workbook.

**Figure 1.11**

The Excel window with the sheet tabs indicated.

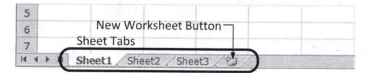

You can move quickly from worksheet to worksheet by selecting a sheet tab. You can also use the arrows to the left of the sheet tabs to move from sheet to sheet, which can be useful when a workbook contains a large number of worksheets. By default, Excel creates three worksheets when you create a new workbook.

The rightmost sheet tab is actually a button that can be used to add a new worksheet to the workbook. The maximum number of worksheets in a workbook is limited only by the amount of memory on your computer.

### 1.4.6 Status Bar

The *Status bar* is normally positioned at the very bottom of the Excel screen. The Status bar displays information about a command in progress and shows some aggregate values for a selected cell range. In Figure 1.12, the Status bar shows that Excel is in **Ready mode** (ready for data entry). When multiple cells are selected, the average, count, and sum of the selected values are displayed in the Status bar. Right-click on the Status bar to customize the display.

## 1.5 GETTING HELP

Excel contains a large online *help system*. To access the help menu, click the Help button on the right side of the Ribbon, as indicated in Figure 1.13. (Excel 2003: choose **Help** from the Menu bar.) The Excel Help window will open, as shown in Figure 1.14.

**Figure 1.12**

The Status bar shows the current data entry mode (Ready), and some aggregate statistics about selected values.

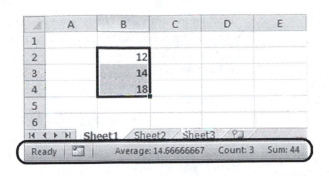

**Figure 1.13**

The Help button is located on the right side of the Ribbon.

**Figure 1.14**

The Excel Help window.

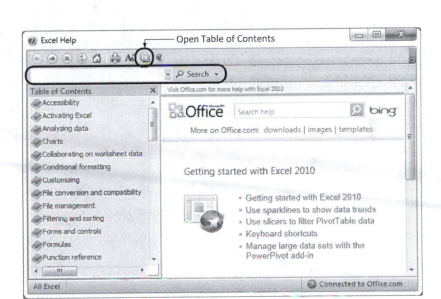

The Help window provides several ways to obtain help, including:

- Browsing the Help Topic List
- Searching the Help system

Each of these methods will be discussed in the next sections.

### 1.5.1 Browsing the Help Topic List

This method is useful if you have time to read about a general topic. Reading through a topic could serve as a tutorial and may provide related information that can expand your skill base, but it is not the method to use if you have a specific question and you want an immediate answer. To view a Help topic, simply select the title in the Browse Excel Help list, or open the Table of Contents using the button indicated in Figure 1.14.

In Excel 2003, open the Table of Contents using these steps:

1. Choose **Help → Microsoft Excel Help** from the Menu bar (or press **F1**). The Help Task pane will be displayed.
2. Click the Table of Contents link on the Task pane.

### 1.5.2 Searching the Help System

While the items in the Table of Contents provide general information about help topics, the quickest way to find answers to specific questions is to search the Excel Help system. Simply type a keyword or a question into the search box, shown in Figure 1.14. (Excel 2003: There are search fields on the Help Task pane and on the Menu bar.)

You enter a keyword or a question in the search field to search the Help system. Figure 1.15 illustrates the result of searching the help system for the word "sine." Notice that the term "sine" was found in three Help topics. Clicking on any of the Help topic titles will cause the topic to be displayed.

**Figure 1.15**

Results from searching for "sine" in the Help system.

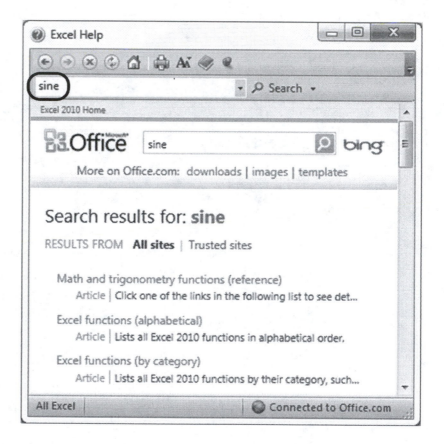

## 1.6 CREATING AND SAVING WORKSHEETS AND WORKBOOKS

### 1.6.1 Creating a New Workbook

When the Excel application is started, a blank workbook containing (by default) three worksheets is automatically created. To create another new workbook, follow these steps:

1. Click the **File** tab to open the File panel shown in Figure 1.16.
2. Click the **New** tab to show options for new workbooks.
3. Click the **Blank Workbook** button to open a new blank workbook.

**Figure 1.16**
The File panel menu with New tab selected

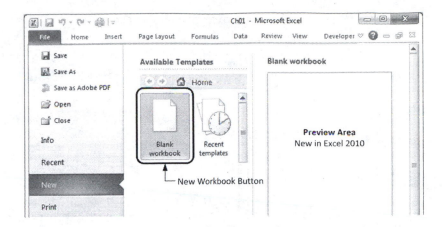

The process is slightly different in older versions of Excel:

- Excel 2007: **Office** button ➔ **New** button ➔ **Blank Workbook** icon ➔ **Create** button
- Excel 2003: Use **File** ➔ **New** and then choose **New** ➔ **Blank Workbook** from the Task pane

### 1.6.2 Opening an Existing Workbook

Like most software packages, Excel keeps a list of the workbooks that you have used recently. To open a recently used workbook, do the following:

1. Click the **File** tab to display the File panel.
2. Click the **Recent** button. The File panel will display a list of recently used workbooks, as illustrated in Figure 1.17.
   - Excel 2007: Click the **Office** button
   - Excel 2003: Choose **File** ➔ **Open** from the Menu bar
3. Click on the file name to open the file in Excel.

**Figure 1.17**
The Office menu with the Open button selected.

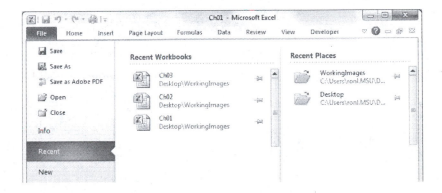

If the workbook does not appear on the list of recently used workbooks, then:

1. Click the **File** tab to display the File panel.
2. Click the **Open** button. The Open dialog box will be displayed (Figure 1.18).

**Figure 1.18**
The Open dialog box.

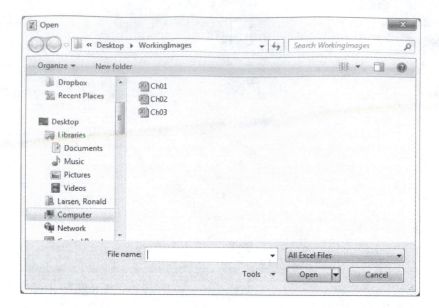

3. Browse for the file you want to open.
4. Click **Open** to open the file in Excel.

For older versions of Excel:

- Excel 2007: **Office** button → **Open** button → Browse for the desired file → **Open** button
- Excel 2003: Choose **File** → **Open** from the Menu bar

From the Open dialog box, you can type in a path and file name in the **File name** field, or you can browse the file system to locate a file. The icons along the left side of the Open dialog box are used to help you find folders quickly.

### 1.6.2.1 Excel File Extensions

Prior to Excel 2007, the *file extension* for an Excel file was .xls. Since Excel 2007, two file extensions are being used:

- .xlsx—the default file name extension, macros disabled.
- .xlsm—macro-enabled workbook.

The .xlsx file name extension indicates that macros (and Visual Basic programs) have been disabled. This ensures that the workbook cannot transmit a macro virus. If the file you want to open uses the .xlsm file extension, macros and Visual Basic programs are enabled and you should only open them if you trust the source.

### 1.6.3 Creating a New Worksheet

Within a workbook, you can have many worksheets. The number of worksheets that you can have in a single workbook is limited only by the available memory on your computer.

To create a new worksheet in an open workbook, click the **Insert Worksheet** button that is the rightmost sheet tab (See Figure 1.19). (Excel 2003: Choose **Insert** → **Worksheet** from the Menu.)

**Figure 1.19**
The Insert Worksheet button on the sheet tab row.

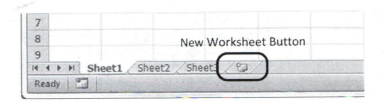

**Figure 1.20**
Using worksheets to organize your work.

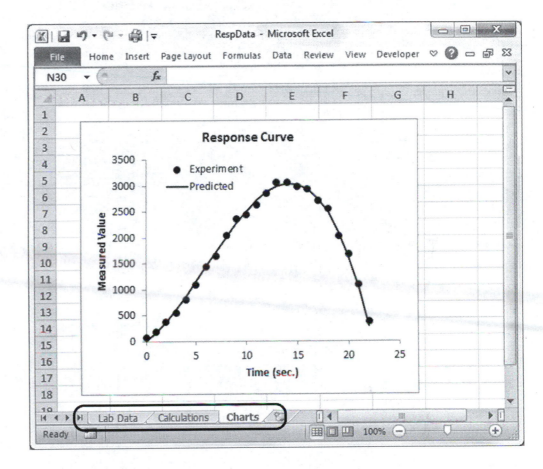

Using multiple worksheets can help keep your work organized. For example, in preparing a laboratory report you might use the following worksheets, as shown in Figure 1.20:

- Lab Data
- Calculations
- Charts

To assign a descriptive name to a worksheet tab:

1. Double-click on the worksheet tab to select the tab and enter text entry mode.
2. Type the new worksheet name.
3. Click anywhere outside the worksheet tab to complete the text entry.

### 1.6.4 Introduction to Templates

A *template* is a workbook that has some of its cells filled in. If you use similar formatting for many documents, then you will benefit from creating and using a template.

You may build your own template or customize preformatted templates and, in time, create a library of your own templates. Excel is installed with a number of sample templates, including one that creates a Loan Amortization Schedule. To open the Loan Amortization template, follow these steps:

**Excel 2010**

1. Click the File tab to display the File panel.
2. Click the New tab to display options for new workbooks, shown in Figure 1.21.
3. Click the **Sample templates** button from the Available Templates list. This will cause a list of sample templates to be displayed on the File panel as shown in Figure 1.22.
4. Select **Loan Amortization** from the Installed Templates list.
5. Click the **Create** button to open the template.

**Figure 1.21**
Choosing an installed template.

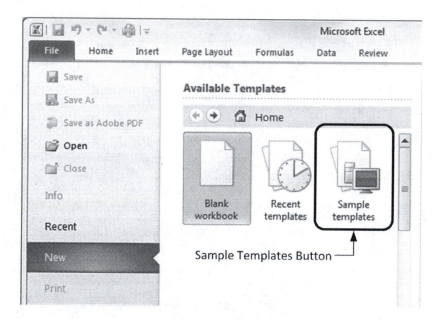

**Figure 1.22**
Selecting the Loan Amortization template.

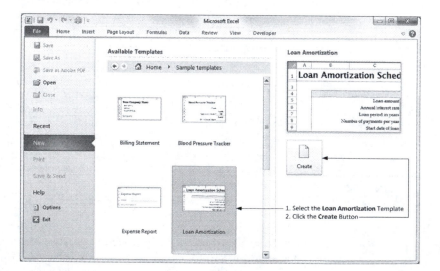

### Excel 2007

1. Click the Office button to open the Office menu.
2. Click the New button to open the New Workbook dialog box, shown in Figure 1.21.
3. Choose **Installed Templates** from the **Templates** list.
4. Select **Loan Amortization** from the **Installed Templates** list.
5. Click the **Create** button to open the template.

### Excel 2003

1. Choose **File ➔ New** from the Menu bar. The New Workbook Task pane will be displayed.
2. Choose **On my Computer . . .** from the **Templates** section. The Templates dialog box will open.
3. Choose the **Spreadsheet Solutions** panel.
4. Select the **Loan Amortization** template.

The resulting Loan Amortization workbook is quite large. Only a portion is shown in Figure 1.23.

**Figure 1.23**
A portion of the Loan Amortization Schedule.

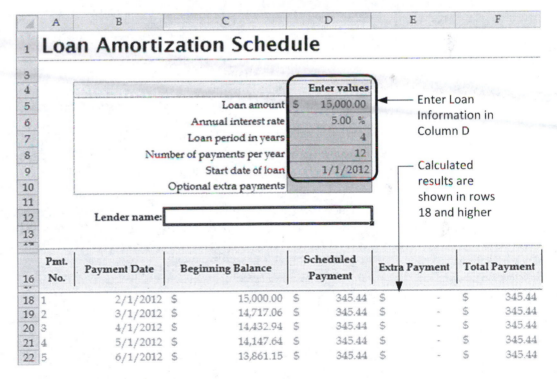

The Loan Amortization template is a preassembled worksheet. Fill in the blank cells labeled:

- Loan amount          $15,000 in this example
- Annual interest rate     5%
- Loan period in years     4 years
- Number of payments per year  12
- Start date of loan       1/1/2009

The worksheet will build an amortization table for you. An amortization table shows a list of required payments on a loan and the amount remaining to be paid after each payment. When all of the required values are entered, the worksheet is automatically completed to show the required payments. The result is shown in Figure 1.23.

### 1.6.5 Opening Workbooks with Macros

A *macro* is a short computer program that records a group of tasks. Excel stores macros in a Visual Basic (programming language) module. By using macros, a set of frequently repeated commands can be stored and then executed with a single mouse click whenever needed.

Macros are very powerful tools. However, macros can contain a *macro virus* that will infect files on your computer. For this reason, you should only enable macros if you are certain of the origin of the macro. If you are unsure of the source of a macro, you should check the document by using virus-protection software before opening the document. Virus-protection software is not provided with Microsoft Excel and must be purchased separately.

Since Excel 2007, there are now two file extensions used with workbooks:

- .xlsx—the default file name extension, macros disabled.
- .xlsm—macro-enabled workbook.

The default .xlsx file name extension tells you that macros (and Visual Basic programs) are disabled. This ensures that the workbook cannot transmit a macro virus. The .xlsm file extension means macros and Visual Basic programs are enabled; you should be careful when opening .xlsm files.

Because of the harm that can be done by macro viruses, Excel comes with Macro Security enabled. To check or change the level of macro security on your installation of Excel, follow these steps:

1. Click the **File** tab to display the File panel (Excel 2007: Office button).
2. Click the **Excel Options** button near the bottom of the options list on the left side of the window. The Excel Options dialog box will open as shown in Figure 1.24.
3. Choose the **Trust Center** panel.
4. Click the **Trust Center Settings . . .** button (shown in Figure 1.24). The Trust Center dialog box will open (shown in Figure 1.25).
5. Click **Macro Settings**. The current level of protection is shown in the Macro Settings list.

In **Excel 2003**, follow these steps:

1. Choose **Tools ➔ Options** from the Menu bar.
2. Choose the **Security** tab.
3. Click the **Macro Security** button. The Security dialog box will open.
4. Choose the **Security Level** tab.

As shown in Figure 1.25, the security has been set so that macros are disabled, but you are notified (and have an option to enable, if desired).

**Figure 1.24**
The Excel Options dialog box, Trust Center panel.

**Figure 1.25**
The Trust Center dialog box showing the current level of macro security.

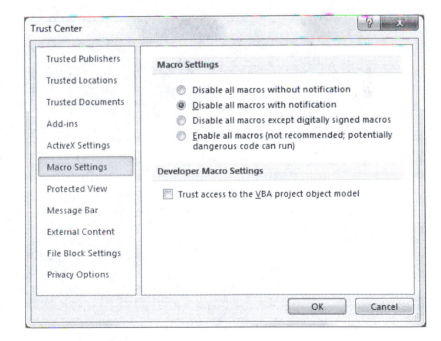

### 1.6.6 Saving Documents

The first time you save an Excel workbook, you need to assign the workbook a name and choose a folder. To save a document for the first time, follow these steps:

**Excel 2010**

1. Click the **File** tab to display the File panel.
2. Click the **Save As . . .** button. The Save As dialog will open as shown in Figure 1.26.
3. Browse for the desired folder to store the workbook.
4. Enter the workbook name in the **File name** field. In this example, "Ch01" was entered as the workbook name. You do not need to enter the file extension. Excel 2010 defaults to the .xlsx file extension, but you can change this using the **Save as type:** drop-down menu to choose another file format.
5. Click **Save** to save the workbook with the entered file name in the selected folder.

**Figure 1.26**
The Save As dialog.

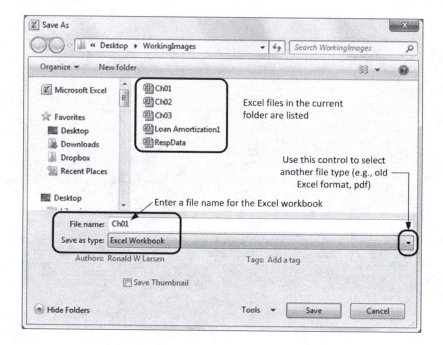

**Excel 2007**

1. Click the **Office** button to open the Office menu.
2. Move the mouse over the **Save As . . .** button. The **Save a copy of the document** options are displayed as shown in Figure 1.26.
3. Select one of the following **Save** options:
   - Excel Workbook (.xlsx)—this is the default format in Excel 2007.
   - Excel Macro-Enabled Workbook (.xlsm)—use only if macros or Visual Basic programs are stored with the workbook.
   - Excel Binary Workbook (.xlsb)—rarely used except for very large workbooks.
   - Excel 97-2003 Workbook (.xls)—used if compatibility with older versions of Excel is needed.
   - Other Formats (e.g., htm)—used to access various less common formats such as .htm for web pages.
4. The Save As dialog will open as shown in **Figure 1.26**.
5. Browse for the desired folder to store the workbook.

6. Enter the workbook name in the **File name** field. In this example, "Ch01" was entered as the workbook name. You do not need to enter the file extension; Excel will automatically add the file extension shown in the **Save as type** field (.xlsx in this example).
7. Click **Save** to save the workbook with the entered file name in the selected folder.

**Excel 2003**

1. Choose **File → Save As** from the Menu bar. The Save As dialog box will open.
2. Browse for the desired folder to store the workbook.
3. Enter the workbook name in the **File name** field.
4. Click **Save** to save the workbook.

To save an open document that was previously saved and assigned a file name:

1. Click the **File** tab to display the File panel.
2. Click the **Save** button to resave the workbook with any changes.

   Or, click the **Save** button on the Quick Access Toolbar.

In older versions of Excel:

- Excel 2007: **Office** button **→ Save** button (or click the **Save** button on the Quick Access Toolbar)
- Excel 2003: **File → Save** from the Menu bar

You should save your work frequently. It is also important to make *backup* copies of your important documents on CDs, or some other physical device. There are many tales of woe from students (and professors) who have lost hours of work after a power failure.

### 1.6.7 The AutoRecover Feature

Excel has an automatic recovery feature, called *AutoRecover*, that can help protect your work from a power failure. When AutoRecover is on, Excel automatically saves a copy of your workbook periodically. Then, if there is a power failure or Excel crashes for any reason, the most recent copy of your workbook can be opened to recover most of your work.

> **Note:** AutoRecover files are erased each time you save your workbook, so using AutoRecover is not equivalent to creating backup copies of your important workbooks. The task of making backup copies is something that you must perform manually.

> To check or change the AutoRecover features, follow this procedure:

1. Click the **File** tab to display the File panel. (Excel 2007: Use the Office button.)
2. Click the **Excel Options** button. The Excel Options dialog will open as shown in Figure 1.24.
3. Choose the **Save** panel (shown in Figure 1.27).
4. If the box next to **Save AutoRecover information** is checked, then the AutoRecover feature is active.
5. Use the **every** field to change the time interval.

**Excel 2003**

1. Choose **Tools → Options** from the Menu bar. The Options dialog box will open.
2. Chose the **Save** tab.

**Figure 1.27**
The Excel Options dialog box, Save panel.

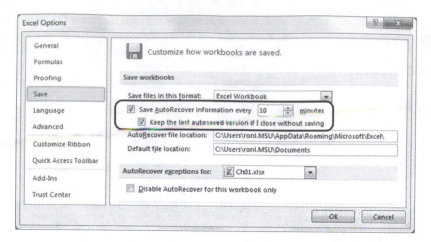

3. If the box next to **Save AutoRecover information** is checked, then the AutoRecover feature is active.
4. Use the **every** field to change the time interval.

While you have the Options dialog box open, take some time to view the other user options that may be customized. Browse through the other tabs on the Options dialog box. Until you become more familiar with Excel, you should probably leave most of the options set to their default values.

### 1.6.8 Naming Documents

It is important to develop a methodical and consistent method for naming worksheets. Over time, the number of worksheets that you maintain will grow larger, and it will become harder to locate or keep track of them. Documents that are related should be grouped together in a separate folder. Do not use the default workbook names (i.e., Book1, Book2, Book3, etc.) or chaos will soon ensue.

If documents are not given meaningful names, then the documents may be inadvertently overwritten. Documents that have very general names (e.g., Workbook) will be difficult to locate later.

One approach that students might use is to create a folder for each course, and use the assignment number with a brief description as the workbook name. In the example shown in Figure 1.28, ENGR 101 might be a computer course, and ENGR 262 a fluid mechanics course.

**Figure 1.28**
Using folders to organize homework files.

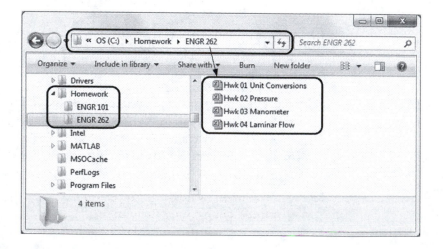

### 1.6.8.1 File Formats and File Extensions

Prior to Excel 2007, the file extension for an Excel file was .xls. Since Excel 2007 new file formats (the way the data is stored) are being used as well as new file extensions (.xlsx and .xlsm). The new file format is called *Office Open XML* and it is intended to improve file management and data recovery. What Excel 2007 and 2010 users need to be aware of is that **workbooks saved in the new format cannot be read in older versions of Excel** (at least, not without a translator). However, workbooks saved in Excel 2003 (or older) can be opened in Excel 2007 and 2012. But the.xlsx file extension offers a higher level of protection from macro viruses, and its use is recommended.

A common scenario during a transition from one version of a program to another is that you may use a new version at school or work, and still have the older version at home (or *vice versa*). As long as you continue to use the older version of Excel, you will need to save your workbooks using the old format. The **Save As** option on the File panel makes this possible. Just be sure to use the Save As type: drop-down and choose to save as Excel 97-2003 Workbook (indicated in Figure 1.29).

**Figure 1.29**
Saving a workbook for older versions of Excel.

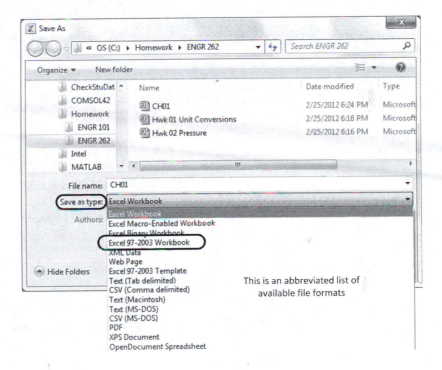

## 1.7 MOVING AROUND A WORKSHEET

There are several methods of moving from place to place in an Excel worksheet. If the worksheet is relatively small, all of these methods will work equally well. As a worksheet grows in size, movement becomes more difficult, and you can save a lot of time by learning the various movement methods.

The currently selected cell is called the *active cell*, and the cell name (e.g., D3) is displayed in the *Name box* on the left-hand side of the Formula bar, as shown in Figure 1.30.

The three general methods for moving around a document are as follows:

- Movement by using the keyboard
- Movement by using the mouse
- Movement by using the Go To dialog box

**Figure 1.30**
The active cell (D3) is
identified in the Name Box.

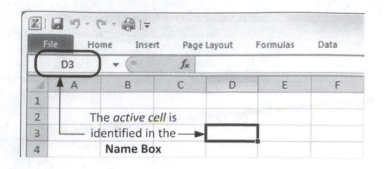

## 1.7.1 Movement by Using the Keyboard

The keyboard may be used to select a worksheet from a workbook. The keyboard may also be used to navigate around a single worksheet quickly and effectively. You may already use the arrow keys to move up, down, left, and right. Combining the Ctrl key with the arrow keys gives you the means for rapid movement. Table 1.1 lists the most frequently used key combinations for movement.

**Table 1.1 Movement Using the Keyboard**

| Key Combination | Action |
| --- | --- |
| ← | Move one cell to the left |
| → | Move one cell to the right |
| ↑ | Move up one cell |
| ↓ | Move down one cell |
| Ctrl + → | Move to the far right of the worksheet |
| Ctrl + ↓ | Move to the bottom of the worksheet |
| Page Down | Move down one screen |
| Page Up | Move up one screen |
| Ctrl + Page Down | Select next worksheet |
| Ctrl + Page Up | Select previous worksheet |
| Home | Move to far-left column of worksheet |
| Ctrl + Home | Move to top-left cell of worksheet (A1) |
| End, → | Move to right end of contiguously filled cell range |
| End, ↑ | Move to top of contiguously filled cell range |
| End, ← | Move to left end of contiguously filled cell range |
| End, ↓ | Move to bottom of contiguously filled cell range |

## PRACTICE!

1. Open a new workbook.
2. Create several worksheets in the workbook using the **Insert Worksheet** button on the sheet tab bar. (Excel 2003: **Insert → Worksheet**.)

3. Create a block of cells containing values, as shown in Figure 1.31.
4. Practice the keyboard movement commands in Table 1.1.
5. Move to the far right and bottom row of a worksheet. What is the maximum size of a worksheet?

| ◢ | A | B | C | D | E | F |
|---|---|---|---|---|---|---|
| 1 | | | | | | |
| 2 | | 1 | 3 | 5 | 7 | |
| 3 | | 2 | 4 | 6 | 8 | |
| 4 | | 3 | 5 | 7 | 9 | |
| 5 | | 4 | 6 | 8 | 10 | |
| 6 | | 5 | 7 | 9 | 11 | |
| 7 | | | | | | |

**Figure 1.31**
A 5 × 4 block of contiguously filled cells for experimenting with the End key movements.

ANSWER
A worksheet is 1,048,576 rows by XFD (16,384) columns in Excel 2007; 65,536 rows × 256 columns in Excel 2003.

### 1.7.2 Movement by Using the Mouse

The mouse is the most common way to move within a worksheet, at least fairly small worksheets. To select a worksheet, choose a tab from the sheet tab bar as indicated in Figure 1.32.

**Figure 1.32**
Click on a sheet tab to display that worksheet.

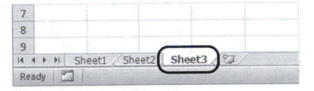

One method of moving around a worksheet with the mouse is to click on a cell. This is most useful if the new insertion point is located on the same screen. If the desired location is on a different page, then the Vertical and Horizontal scroll bars may be used to move quickly to a distant location.

### 1.7.3 Movement by Using the Go To Dialog Box

If you have a large worksheet that covers many screens, then using the keyboard and mouse can be a cumbersome way of moving through the worksheet. The Go To dialog box offers a method for moving directly to distant locations on the worksheets.

To move to a location using the Go To feature, do the following:

1. Open the Go To dialog box with Ribbon options: **Home tab → Editing group → Find & Select drop-down menu → Go To . . . button**. (Excel 2003: **Edit → Go To**.) The Go To dialog box will open, as depicted in Figure 1.33.

Or, you can press the **F5** key to open the Go To dialog box.

2. Type in a *cell reference*. For example, type G36 and then click **OK**.

**Figure 1.33**
The Go To dialog box.

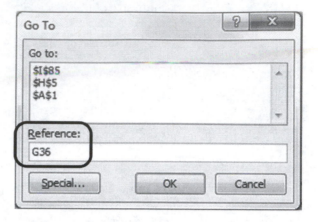

The screen will display cell G36, and it will become the active cell.

A history of previous references is kept in the **Go To** window, so recently visited cells can be quickly located simply by selecting them with the mouse.

In addition to moving to cells by location, you can move to cells of a particular type. We have not yet shown you how to create cells of different types. However, imagine that you have created a number of cells containing formulas. You can locate formulas with errors in them by using the Go To Special dialog box as follows:

1. Open the Go To Special dialog box with Ribbon options: **Home tab → Editing group → Find & Select drop-down menu → Go To Special . . . button**. The Go To Special dialog box will open, as depicted in Figure 1.34.

**Figure 1.34**
The Go To Special dialog box.

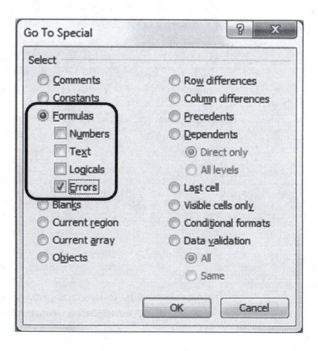

Or, you can click the **Special . . .** button on the Go To dialog box.

2. Select the type of cell you want to locate (e.g., **Formulas** with **Errors**) and then click **OK.**

The first formula with an error will become the active cell, and all other formulas with errors will be highlighted.

## 1.8 SELECTING A REGION

Much of the time spent in worksheet preparation involves moving, copying, and deleting regions of cells or other objects. In this section, we will be selecting regions of cells, but the same principles apply to regions that contain charts, formulas, and other objects. Before an action can be applied to a region, the region must be selected. The selection process can be performed by using either the mouse or the keyboard.

### 1.8.1 Selection by Using Cell References

In many cases, you will have the option of typing a cell reference. For example, you can type cell references into a formula. A single cell is noted by its column letter and row number. A rectangular range of cells is denoted by the reference for the top-left and bottom-right cells. For example, the rectangle bordered by B2 on the top left and E5 on the bottom right is denoted as B2:E5 (see Figure 1.35). Note that the first selected cell (cell B2 in Figure 1.35) is shown in a different color, and indicates the active cell.

**Figure 1.35**
The selected cell range B2:E5.

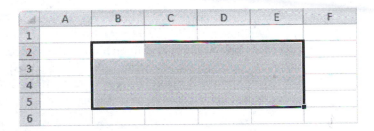

### 1.8.2 Selection by Using the Mouse

To select a region of cells, called a *cell range*, with the mouse, click the mouse on the first cell in the range. Then drag the mouse cursor to the cell at the other end of the range. As you drag the mouse, the selected region will be highlighted.

To select a cell range that is larger than one screen, drag the mouse to the bottom of the screen. If you hold the mouse at the bottom of the screen without releasing the mouse button, the screen will scroll and the selected region will continue to grow. This takes a little practice.

To select a whole column, click on the column header. To select a whole row, click on the row header. This is illustrated in Figure 1.36.

To select the entire worksheet, choose the header at the top-left corner of the worksheet, between A and 1, as illustrated in Figure 1.37. This unlabeled header is called the **Select All** button. This is useful if you are applying a change to every cell in a worksheet.

**Figure 1.36**
Selecting an entire row.

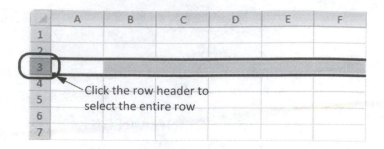

Click the row header to select the entire row

**Figure 1.37**
The Select All button.

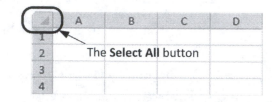

The **Select All** button

### 1.8.3 Selection by Using the Keyboard

An alternative method for selecting regions of a document is to use the keyboard, as follows:

1. Click the mouse on one corner of the region that you wish to select.
2. Hold down the **Shift** key and use the arrow keys to move to the other end of the region.
3. Release the **Shift** key.

The selected region will be highlighted. If you make a mistake and incorrectly select a region, then click the mouse cursor anywhere on the worksheet window before you apply an action (such as delete). If the highlighting disappears, then you have deselected the region.

## PRACTICE!

Try the following exercise to practice selecting regions:

1. Click in cell B2 and type the number 5.
2. Press the **down-arrow** key.
3. Type the number 6.
4. Press the **down-arrow** key.
5. Type the number 7.
6. With the mouse, select cell range B2:B4, as shown in Figure 1.38.
7. Choose the **SUM** button on the Ribbon's Home tab: **Home tab →
   Editing group → SUM button**. (Excel 2003: **AutoSum** button on the
   Standard toolbar.)

A formula for cell B5 will be added that contains the sum of cells B2, B3, and B4. The results should resemble Figure 1.39.

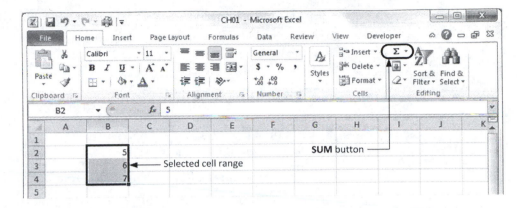

**Figure 1.38**
Click the SUM button after selecting the cells to be added.

**Figure 1.39**
The *SUM* function is entered just below the selected cell range.

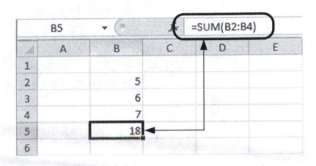

## 1.9 CUTTING, MOVING, COPYING, AND PASTING

Once a region has been selected, several actions may be taken, such as delete, move, copy, and paste. As usual, Excel provides several ways to accomplish the same actions. These include using keyboard commands and mouse commands.

The cut, copy, and paste commands make use of a special location called the *Windows clipboard*. The clipboard is a temporary storage location that can be used to hold the contents of a cell, a range of cells, or most other objects such as charts. To view the contents of the clipboard, click the **Clipboard** or **Expand** button at the bottom-right corner of the Clipboard group in the Ribbon's Home tab as shown in Figure 1.40. (You do not need to see the clipboard contents to use the clipboard.)

**Figure 1.40**
The Clipboard group on the Ribbon's Home tab.

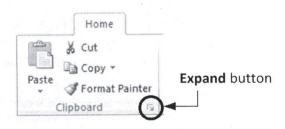

### 1.9.1 Cutting a Region

*Cutting* a region (e.g., a range of cells) removes the contents of the selected region from the worksheet and leaves them on the clipboard. A region may be cut by using the mouse or the keyboard.

To cut a region using the mouse, follow these steps:

1. Select a region.
2. Click the **Cut** button in the Clipboard group in the Ribbon's Home tab. (Excel 2003: Choose **Edit → Cut**.)

The region to be cut will be highlighted by a rotating dashed line.

Alternative methods for cutting a selected region include the following:

- Select the region to be cut and then right-click on the selected region. Select Cut from the pop-up menu.
- Select the region to be cut and then press **Ctrl + X**.

No matter which method you use to cut the region, the effect is to place the contents of the region on the clipboard. This will be displayed in the Clipboard Task pane, if the pane is visible. Figure 1.41 illustrates a region of four cells in Column B that have been selected and cut.

**Figure 1.41**
Four cells on the clipboard.

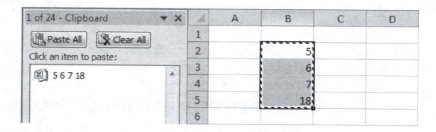

Notice that the cut cells have not been removed from the worksheet. The process of cutting the cells marks the cells for removal, but they are not actually removed unless the cut procedure is followed by a paste procedure. This is described in the next section.

### 1.9.2 Moving a Region (Cut and Paste)

A region may be moved by first cutting the region (to the clipboard) and then pasting it (from the clipboard) to the new location. The cut and paste operation may be performed by using the mouse or the keyboard.

To move a region using the mouse, do the following:

1. Select and cut a region. This places the contents of the region on the clipboard.
2. Select a destination cell or region.
3. Click the **Paste** button in the Clipboard group in the Ribbon's Home tab. (Excel 2003: Choose **Edit → Paste** from the Menu bar.)

The region of cells should now appear in the new location. If you do not select a destination region of the same size and shape as the cut region, then Excel will create a region with the appropriate size.

Alternative methods for pasting clipboard contents include the following:

- Right-click on the selected destination region, and then select **Paste** from the pop-up menu.
- Select the destination. Then press **Ctrl + V**.

### 1.9.3 Copying a Region

*Copying* a region is very similar to moving a region, except that the contents of the original region remain intact; they are copied to the clipboard but not cut (moved) to the clipboard.

To copy a region using the mouse, follow these steps:

- Select a region.
- Click the **Copy** button in the Clipboard group in the Ribbon's Home tab. (Excel 2003: Choose **Edit → Copy**.)

The region to be copied will be highlighted by a rotating dashed line.

Alternative methods for copying a selected region include the following:

- Select the region to be copied then right-click on the selected region. Select **Copy** from the pop-up menu.
- Select the region to be copied, then press **Ctrl + C**.

**Note:** The keyboard shortcuts for cutting (**Ctrl + X**), copying (**Ctrl + C**), and pasting (**Ctrl + V**) use adjacent keys to make them easier to remember (Figure 1.42).

**Figure 1.42**
Cut (X), Copy (C), and Paste (V) keyboard shortcuts.

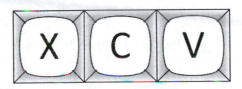

## 1.10 INSERTING AND DELETING CELLS

New cells may be added to a worksheet, and existing cells may be deleted (removed) or cleared (emptied).

### 1.10.1 Deleting Cells

Deleting a region of cells removes the cells from the worksheet. The vacancies, or holes, that are left behind must be filled in, and Excel will open the Delete dialog box (shown in Figure 1.43) to ask you how you want to fill the vacancies.

To delete a region of cells, follow these steps:

1. Select the region of cells to be deleted.
2. Right-click the selected region and choose **Delete . . .** from the pop-up menu. The Delete dialog box will appear, as shown in Figure 1.43.
3. Choose whether you want Excel to fill the vacancies created by deleting the cells by:
   - shifting the remaining cells up or to the left,
   - shifting the entire row below the vacancies up, or
   - shifting the entire column to the right of the vacancies to the left.
4. Click **OK** to close the Delete dialog box and delete the selected cells.

**Figure 1.43**
The Delete dialog box.

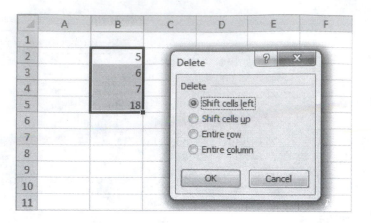

## 1.10.2 Clearing Cells

To remove the contents of cells without deleting the cells themselves, perform these steps:

1. Select the region of cells to be deleted.
2. Right-click the selected region and choose **Clear Contents** from the pop-up menu (or, press the **Delete** key).

## 1.10.3 Inserting Cells

You can insert new cells, rows, columns, or an entire worksheet using the **Insert drop-down menu** on the Ribbon's Home tab (see Figure 1.44): **Home tab → Cells group → Insert drop-down menu**. (Excel 2003: Use the Insert menu option.)

**Figure 1.44**
The Insert drop-down menu.

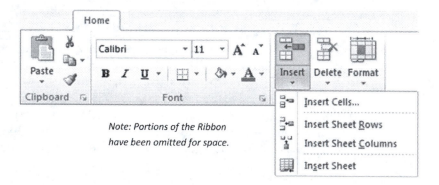

*Note: Portions of the Ribbon have been omitted for space.*

## 1.11 SHORTCUT KEYS

As a novice user, you may have trouble finding commands. The Ribbon in Excel 2007 and 2010 has been designed to get commonly used commands where you can find them, but it still takes some getting used to. *Shortcut keys* are the quickest way to execute a command and can save time, but they have to be memorized. The good news is that most are commonly used by lots of programs, not just Excel. Table 1.2 lists common shortcut key combinations.

**Table 1.2 Commonly Used Shortcut Keys**

| Command | Shortcut |
|---|---|
| New Workbook | Ctrl + N |
| Open Workbook | Ctrl + O |
| Save Workbook | Ctrl + S |
| Print | Ctrl + P |
| Undo | Ctrl + Z |
| Cut | Ctrl + X |
| Copy | Ctrl + C |
| Paste | Ctrl + V |
| Find | Ctrl + F |
| Replace | Ctrl + H |
| Go To | Ctrl + G |
| Format Cells | Ctrl + 1 |
| Help | F1 |
| Spell Check | F7 |

One method of learning some of the shortcuts is to look at the *Screen Tips* for Ribbon items. Screen Tips are descriptions that are displayed when you let the mouse hover over a Ribbon item. For example, in Figure 1.45, the Screen Tip for the Copy button is shown, and it indicates that the keyboard shortcut for the copy operation is **Ctrl + C**.

**Figure 1.45**
The Screen Tips for Ribbon items often indicate the keyboard shortcut.

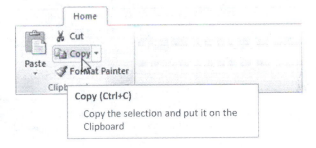

## 1.12 FINDING AND CORRECTING MISTAKES

Let's face it, mistakes happen. Finding mistakes in a complex Excel worksheet can be a challenge. A couple of simple fixes are described here:

- Undo (**Ctrl + Z**)
- *Spell Check* (**F7**)
- AutoCorrect

### 1.12.1 Undoing Mistakes

Excel allows actions to be undone or reversed. To undo the last action, click the **Undo** button on the Quick Access Toolbar (indicated in Figure 1.46), or type **Ctrl + Z**. (Excel 2003: Choose **Edit → Undo** from the Menu bar.)

**Figure 1.46**
The Undo button on the
Quick Access Toolbar.

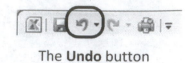

The **Undo** button

To see the list of recent actions, choose the down-arrow button next to the **Undo** button. From this list, you may select one or more actions to be undone. Note that if you select an action on the list, then all of the actions above it in the list will also be undone! If you accidentally undo an action, then you may redo it by selecting the **Redo** button, which is next to the **Undo** button.

### 1.12.2 Checking Spelling

Excel can check the spelling of cells containing text. To check the spelling in a region, first select the region, then click the Spelling button on the Ribbon's Review tab: **Review tab → Proofing group → Spelling button** (indicated in Figure 1.47). Or, press the **F7** key. If Excel finds a spelling mistake, then the Spelling dialog box will appear, as shown in Figure 1.48.

**Figure 1.47**
The Spelling button on
the Review tab.

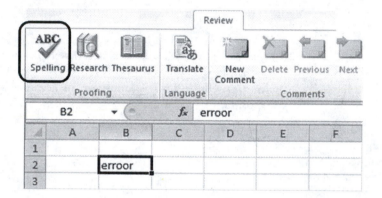

**Figure 1.48**
The Spelling dialog box.

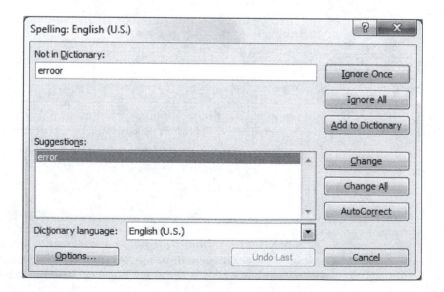

The text thought to be in error is displayed in the top text box. Suggestions for changes are presented in the bottom text box. At any point in the process, you can choose whether to accept or ignore the suggestions. If you choose a suggested correction, then you may click the Change All button to change all occurrences of the misspelled word in the selected region.

You may add new words to the dictionary by choosing the **Add to Dictionary** button. This will probably be necessary as you proceed through your coursework, since many engineering terms are not in the default dictionary.

### 1.12.3 The AutoCorrect Feature

The Excel *AutoCorrect* feature recognizes some spelling errors and corrects them automatically. AutoCorrect performs actions such as automatically capitalizing the first letter of a sentence or correcting a word whose first two letters are capitalized.

You can test to see if the AutoCorrect feature is turned on for your installation of Excel. Try typing the letters *yuo*, then press the spacebar. Was the word automatically retyped as *you*? If so, then you have AutoCorrect turned on.

To see your AutoCorrect settings and dictionary, use **Office → Excel Options → Proofing tab → AutoCorrect Options.** (Excel 2003: **Tools → AutoCorrect Options.**) The AutoCorrect dialog box will appear, as shown in Figure 1.49.

**Figure 1.49**
The AutoCorrect dialog box.

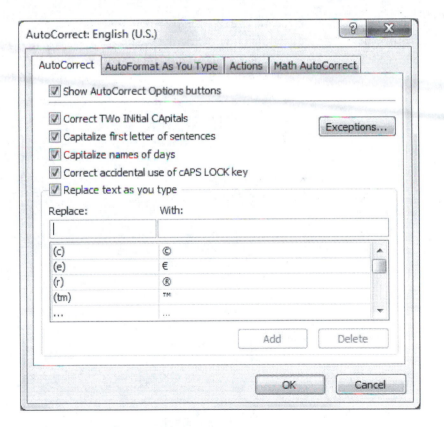

From the AutoCorrect dialog box, you can select (or deselect) various options. You can also scroll through the AutoCorrect dictionary, add entries to the dictionary, and add exceptions to the dictionary. Creating an exception list will be necessary if you use all of the AutoCorrect features. For example, if you have selected the option that automatically converts the second capital letter to lowercase, you may have an occasional exception. Be careful when adding new entries into the AutoCorrect dictionary. You may inadvertently add an entry for a misspelling that is a legitimate word.

## 1.13 PRINTING

Before attempting to *print* a document, make sure that your printer is correctly configured. See your operating system and printer documentation for assistance.

### 1.13.1 Setting the Print Area

An Excel worksheet contains 1,048,576 rows by 16,384 columns. That would be a huge area to print. Excel never prints all cells in a worksheet; it prints a rectangular region that contains all of the cells that have contents. If you want to print a smaller region of a worksheet, you must first set the *print area*. To set the print area, perform the following steps:

1. Select the region that is to be printed.
2. Set the print area using Ribbon options: **Page Layout tab → Page Setup group → Print Area drop-down menu → Set Print Area option**. (Excel 2003: **File → Print Area → Set Print Area**.)

### 1.13.2 Previewing a Worksheet

It is advisable to use the *Print Preview* feature to preview a document before printing it. Many formatting problems can be resolved during the preview process. To preview the document as it will be printed, do the following:

1. Set the print area (if you want to print only a portion of your work).
2. Activate print preview: **File tab → Print tab**.

In Excel 2010 the Print Preview is displayed on the File panel, as shown in Figure 1.50.

**Figure 1.50**

The Print Preview screen is on the File tab in Excel 2010.

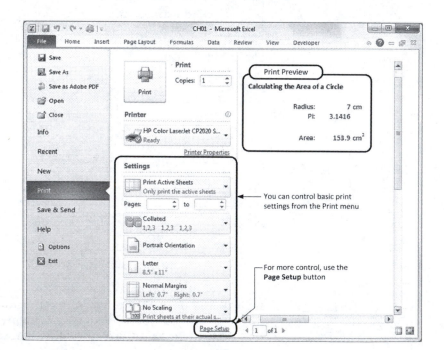

In older versions of Excel, Print and Print Preview are separate operations.

- Excel 2007: **Office button ➔ Print sub-menu ➔ Print Preview**
- Excel 2003: **File ➔ Print Preview**

In Excel 2010 there are a number of useful print adjustments available right on the File panel:

- *Print Active Sheets*—allows you to indicate which pages of the worksheet should be printed.
- *Collated*—if you are printing multiple copies, you can control how they should be stacked.
- *Orientation*—allows you to choose how the worksheet should be printed on the page (portrait or landscape orientation).
- *Page Size*—choose the size of paper to use when printing.
- *Margins*—change the size of the margins to several pre-set options.
- *Scaling*—this is a very useful option that allows you to indicate whether or not the output should be scaled to fit on a certain number of pages.

For more control, click the **Page Setup** link (indicated in Figure 1.50) to open the Page Setup dialog, shown in Figure 1.51. (In older versions of Excel, you will need to open the Page Setup dialog to control most output settings.) Two of the most useful controls (for Excel 2003 and 2007 users) are on the Page panel, shown in Figure 1.51.

- Select **Orientation**: Portrait or Landscape.
- **Fit to** 1 page wide by 1 tall.

The **Fit to** option takes everything that is going to be printed and scales it to fit on the number of pages you indicate. The most common use is to force a worksheet to

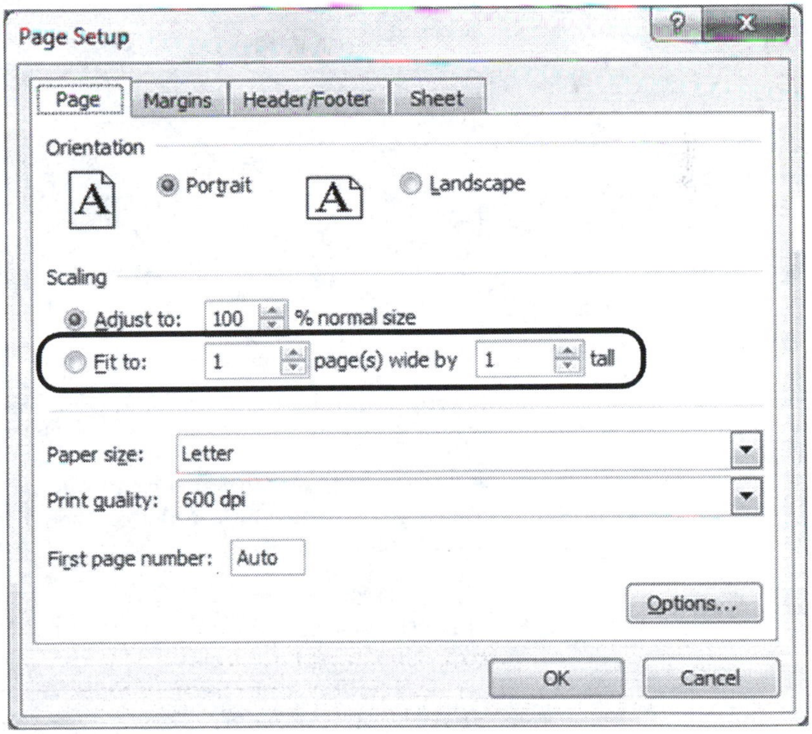

**Figure 1.51**
The Page Setup dialog box, Page panel.

print onto one page. Excel 2010 users can use either the **Scaling** option on the output Settings (on the File panel) or the Page Setup dialog to accomplish this.

The **Margins panel** on the Page Setup dialog box provides another way to adjust margins. The **Header/Footer panel** allows you to print a header or footer on each page of the printout. Options include page numbers, author name, file name, or custom text.

The **Sheet panel** (shown in Figure 1.52) can be used to include:

- **Gridlines** (to show the cells)
- **Row and column headings**

on the printout.

**Figure 1.52**
The Page Setup dialog's sheet tab is useful when gridlines and headings should be printed.

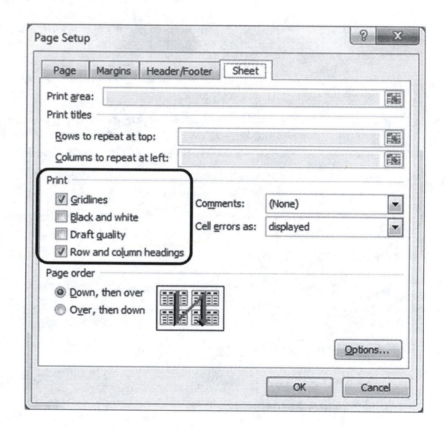

### 1.13.3 Printing a Worksheet

You can print a worksheet in several ways. To print a worksheet, choose one of the following methods:

- Use **File tab → Print tab → Print button.**
  - Excel 2007: **Office button → Print sub-menu → Print**
  - Excel 2003: **File → Print**
- Press **Ctrl + P.**

In Excel 2010, clicking the Print button sends the output to the currently selected printer. In older versions, a Print dialog appears that allows you to select a printer and types of collation. In Excel 2010, these features are included in the File panel when the Print tab is selected.

## KEY TERMS

active cell
AutoCorrect
AutoRecover
backup
cell
cell range
cell reference (e.g., B2)
clipboard (Windows clipboard)
Close button
column heading
control buttons
copy
cut
dialog box
Excel
file extensions (.xls, .xlsx, .xlsm)
formula (equation)

Formula bar
gridlines
group (Ribbon group)
Help System
Home tab
Insert Function button
macro
macro virus
Maximize/Restore button
Minimize button
Name box
Office button
paste
*PI*
print
print area
print preview
Quick Access Toolbar

range (cell range)
Redo button
Ribbon
Ribbon tabs
row heading
Screen Tip
search
sheet tab
shortcut keys
spell check
spreadsheet
Status bar
template
Title bar
Undo button
work area
workbook
workbook window
worksheet

## SUMMARY

### Excel Screen Layout

- Title bar
- Ribbon
- Quick Access Toolbar
- Formula bar
- Work area
- Sheet tabs
- Status bar

### File Tab (Office Button in Excel 2007)

- Open workbooks
- Save workbooks
- Print workbooks
- Set Excel Options

### Control Buttons

- Minimize button
- Maximize/Restore Window Toggle
- Close button

**Ribbon**

Tab → Group → Drop-down Menu → Button

- *File tab*—commonly used to open, save, and print workbooks, and change Excel Options.
- *Home tab*—commonly used commands for formatting and sorting.
- *Insert tab*—used to insert objects such as charts and hyperlinks.
- *Page Layout tab*—used to modify entire sheets (apply themes, set print area, etc.).
- *Formulas tab*—used to insert functions and manage defined names of cells and cell ranges.
- *Data tab*—provides access to sorting and filtering features, and data analysis tools (if activated).
- *Review tab*—used to add comments and track changes to a worksheet.
- *View tab*—used to change the display magnification (zoom), and to show or hide features such as the Formula bar and gridlines.

**Help System (F1)**

- Browsing the Help Topic List
- Searching the Help system

**Working with Excel Workbooks**

- Create a New Workbook: **File tab → New → Blank Workbook**
- Open an Existing Workbook: **File tab → Open → (browse to find file) → Open**
- Saving a Workbook:
    - First time: **File tab → Save As tab → (browse for folder; assign file name) → Save**
    - If already named: **File tab → Save** (or, click **Save** button on Quick Access Toolbar).

**Adding a Worksheet to a Workbook**

- Click the **Insert Worksheet** button that is the rightmost sheet tab.

**Excel File Extensions**

- .xls—version 2003 or earlier
- .xlsx—the default file name extension since Excel 2007, macros disabled
- .xlsm—Excel 2007, 2010 macro-enabled workbook

**Moving Around the Worksheet Using the Keyboard**

| Key Combination | Action |
| --- | --- |
| ← | Move one cell to the left |
| → | Move one cell to the right |
| ↑ | Move up one cell |
| ↓ | Move down one cell |
| Ctrl + → | Move to the far right of the worksheet |
| Ctrl + ↓ | Move to the bottom of the worksheet |
| Page Down | Move down one screen |
| Page Up | Move up one screen |

| | |
|---|---|
| Ctrl + Page Down | Select next worksheet |
| Ctrl + Page Up | Select previous worksheet |
| Home | Move to far-left column of worksheet |
| Ctrl + Home | Move to top-left cell of worksheet (A1) |
| End, → | Move to right end of contiguously filled cell range |
| End, ↑ | Move to top of contiguously filled cell range |
| End, ← | Move to left end of contiguously filled cell range |
| End, ↓ | Move to bottom of contiguously filled cell range |

## Cut or Copy

1. Select a region.
2. Click the **Cut** or **Copy** button in the Clipboard group in the Ribbon's Home tab.

## Paste

1. Select a destination cell or region.
2. Click the **Paste** button in the Clipboard group in the Ribbon's Home tab.

**Shortcut Keys**

| Command | Shortcut |
|---|---|
| New Workbook | Ctrl + N |
| Open Workbook | Ctrl + O |
| Save Workbook | Ctrl + S |
| Print | Ctrl + P |
| Undo | Ctrl + Z |
| Cut | Ctrl + X |
| Copy | Ctrl + C |
| Paste | Ctrl + V |
| Find | Ctrl + F |
| Replace | Ctrl + H |
| Go To | Ctrl + G |
| Format Cells | Ctrl + 1 |
| Help | F1 |
| Spell Check | F7 |

## Printing

### Set Print Area

1. Select the region that is to be printed.
2. Set the print area using Ribbon options: **Page Layout tab → Page Setup group → Print Area drop-down menu → Set Print Area option**

*Print Preview*

- **File tab** → **Print tab**

*Print Alternatives*

- Use **File tab** → **Print tab** → **Print**
- Press **Ctrl + P**

## PROBLEMS

1. Test your understanding by filling in the blanks.
   - The ____ displays the name of the currently open workbook.
   - The Home, Insert, and Page Layout tabs are found on the ____.
   - Clicking on the Save button on the Quick Access Toolbar has the same effect as choosing ____ from the File tab (Office menu in Excel 2007).

2. What is the maximum number of rows and columns for a single Excel worksheet?

3. Use the Insert Function dialog box to identify the Excel function names for the following mathematical functions:
   ____ sine
   ____ arithmetic mean
   ____ natural logarithm
   ____ convert degrees to radians
   ____ remove or truncate the decimal part of a number
   ____ returns e raised to the power of a number

4. Name three ways to undo a mistake.

5. Identify the shortcut keys for the following actions:
   ____ Help
   ____ Copy selected region
   ____ Cut selected region
   ____ Move to the beginning of a worksheet

6. Visit the U.S. National Institute of Science and Technology (NIST) Physics Laboratory's website about the International System of Units (SI) at http://physics.nist.gov/cuu/units.

   Click on the menu item labeled SI units, and locate the table for SI Base Units. Use that table to fill in the missing entries in Table 1.3.

7. The electronic spreadsheet has played an important role in the history of computing. The links presented here discuss the history of electronic spreadsheets. Access these websites with your web browser and then answer the questions that follow:

   Power, Daniel. A Brief History of Spreadsheets, at URL http://dssresources .com/history/sshistory.html.
   Mattessich, Richard. Early History of the Spreadsheet, at URL http://www .j-walk.com/ss/history/spreadsh.htm.

**Table 1.3 SI Base Units**

| Quantity | Name | Symbol |
|---|---|---|
| length | | m |
| | kilogram | kg |
| time | second | |
| electric current | ampere | |
| temperature | | K |
| | mole | mol |
| luminous intensity | | cd |

- What is the name of the first marketed electronic spreadsheet that was partly responsible for the early success of the Apple computer?
- In what year was Excel originally introduced (for Macintosh computers)?

8. Excel's trigonometric function *PI* returns an approximation of the mathematical constant $\pi$. Read the information about *PI* on the Insert Function dialog box to determine the number of digits of accuracy of the constant returned by this function.

9. Describe the difference between three of Excel's logarithm functions: LN, LOG, and LOG10. Use the Help system to find the answer to this question.

10. Explain the difference between Cut-and-Paste and Copy-and-Paste. Which would you use if you needed to:

- move a column of values to a new location within a worksheet?
- create a table in a Word document from a table of values in an Excel worksheet (leaving the Excel worksheet unchanged)?

11. Access Microsoft's website (www.microsoft.com) to find a calendar template for Excel. (Enter *Excel calendar template* in the search box on the Microsoft web page.) How many Excel calendar templates are available for downloading?

12. Perform a Google® search on the phrase *Excel Tips*. On a scale from 0 (no information) to 10 (massive amounts of information), how much information is available about Excel on the World Wide Web?

CHAPTER

# 2

# Entering and Formatting Data

## Sections

## Objectives

*After reading this chapter, you should be able to perform the following tasks:*

- Enter numeric, text, and time and date data into a worksheet.
- Quickly enter series of data by using Excel's Fill Handles.
- Format cells, rows, and columns.
- Password protect your worksheets.
- Format tables.
- Perform sorts on tabulated data.
- Apply conditional formatting to a range of cells.

## 2.1 INTRODUCTION TO ENTERING AND FORMATTING DATA

*Worksheet* cells can be filled with numeric values, text, times, dates, logical values, and formulas. In addition, a cell may contain an error value if Excel cannot evaluate its contents. The most common type of data entered into a spreadsheet is numeric data.

You can control the appearance of entered data by modifying the *formats* applied to the cell containing the data. Formatting of numeric data, for example, allows you to specify the number of displayed decimal places and whether or not to show the value by using scientific notation. Formatting also allows you to set the background and font colors, and to specify the type of border for any cell. Collections of formatting options are called *styles* and Excel 2010 comes with many predefined styles.

Any time you enter data into a cell, it automatically gets a default format (usually *General format*) and a default style, called the *Normal style*. In most situations the default formatting is adequate, so you will not often need to make formatting changes in your worksheets. However, in this chapter the various formatting options will be presented so that you can control the formatting of your worksheet contents when necessary.

Here's a list and the definitions of the most common types of formatting commands you may use to properly represent your data in Excel:

- *General format*—Excel will choose a format on the basis of the contents of the cell. It is the default format for cells containing numbers or text.
- *Currency format*—Excel will display the numeric data with a currency symbol.
- *Accounting format*—Excel will display the numeric data with a currency symbol aligned at the left side of the cell.
- *Date format*—Excel will display the data as a date.
- *Scientific format*—Excel will display the number in scientific notation.
- *Text format*—The contents of the cell will be treated as text.
- *Time format*—Excel will display the data as a time.

## 2.2 ENTERING DATA

A typical Excel worksheet has some cells that contain data that have been entered from a keyboard. Other cells contain formulas that use the entered data to calculate results, which are then displayed on the worksheet. The usual first step in building a worksheet is entering data.

### 2.2.1 Entering Numeric Data

Numbers can be typed into cells by using a keyboard, or imported from another source such as another Excel worksheet, a Word *table*, or a column of values from a website. Numeric values are stored internally with up to 15 digits of precision (including the decimal point).

#### 2.2.1.1 How Excel Determines That a Cell Contains a Number

Each time you enter anything into a cell, Excel checks to see if the new entry appears to be a number, a date, a time, or a formula (an equation). If the new entry cannot be interpreted as any of these specific types, then Excel treats the new entry as text.

Excel will treat a new cell entry as a number if it contains only the following characters:

0 1 2 3 4 5 6 7 8 9

+ − ( ) , . /

$ %

E e

When Excel determines that a cell contains a simple number, it automatically applies the General format. If the $ is included, a *Currency format* is applied; and if the % symbol is included, a Percentage format is applied. In most cases the default formatting is adequate; however, you can change the applied format when necessary. A list of commonly used cell formats is available as a drop-down list on the *Ribbon*'s *Home tab* [**Home tab → Number group**], as shown in Figure 2.1.

**Figure 2.1**
Selecting cell number formats from the Ribbon uses a drop-down menu in the Number group in the home tab.

The available format options on the drop-down list are shown in Figure 2.2.

**Figure 2.2**
Format options.

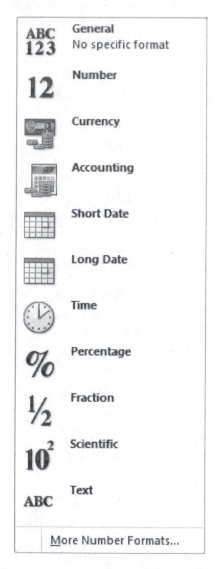

As you can see in Figure 2.2, Excel provides two formats for monetary values: Currency and Accounting, and quick access to two date formats: Short Date and Long Date. Both of the monetary value formats include a currency symbol ($ in the

United States, € in most of Europe), but arrange the values slightly differently, shown in Figure 2.3.

**Figure 2.3**
Comparing the Currency and Accounting formats.

| ◢ | A | B | C | D | E |
|---|---|---|---|---|---|
| 1 | | | | | |
| 2 | | $10.45 | | Currency Format | |
| 3 | | $100.57 | | | |
| 4 | | $10,000.63 | | | |
| 5 | | | | | |
| 6 | | | | | |
| 7 | | $        10.45 | | Accounting Format | |
| 8 | | $      100.57 | | | |
| 9 | | $  10,000.63 | | | |
| 10 | | | | | |

Aligning the currency symbols by using the *Accounting format* makes columns of monetary values easier to read.

The Short Date and Long Date format options shown in Figure 2.14 both apply the Date format to the cells, but display dates with different amounts of information, as shown in Figure 2.4.

**Figure 2.4**
Comparing date formats.

| ◢ | A | B | C | D | E |
|---|---|---|---|---|---|
| 1 | | | | | |
| 2 | | Thursday, February 16, 2012 | | Long Date | |
| 3 | | | | | |
| 4 | | 2/16/2012 | | Short Date | |
| 5 | | | | | |

For additional options, use the [**More Number Formats . . .**] option at the bottom of the drop-down list (shown in Figure 2.2) to open the Format Cells dialog (Figure 2.5). [Excel 2003: Use menu options **Format ➔ Cells.**]

**Figure 2.5**
Format Cells dialog.

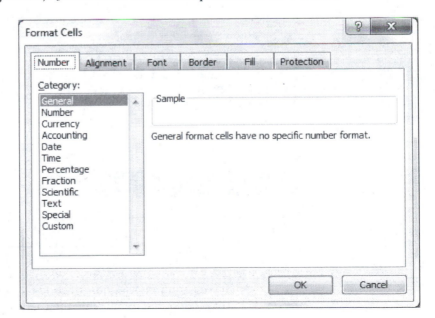

Alternatively, you can open the Format Cells dialog by right-clicking on a cell and choosing **Format Cells** from the Quick Edit menu, as shown in Figure 2.6.

**Figure 2.6**
Using the Quick Edit menu.

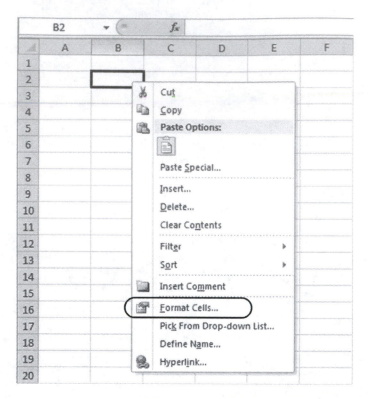

When you change the format applied to a cell, the internal representation of the number stored in the cell is not changed; only the method for displaying the value is modified.

**Note:** When Excel determines that a cell contains a date, the cell contents are changed to a date-time value. This conversion allows Excel to perform math with dates (e.g., determining the number of days between two specified dates).

The General format is Excel's default format. When the General format is used, Excel will apply a format based on the contents of the cell. For example, if you type $3.4 into a cell, Excel will assume that you are entering a monetary value and will automatically convert to Currency format.

## PRACTICE!

Open a blank worksheet, then follow these steps:

1. Enter the values shown in Figure 2.7 into separate cells. Be sure to include the commas, dollar sign, and percentage symbol as shown.
2. Use the format drop-down display (see Figure 2.7) on the Ribbon [**Home tab → Number group**] to determine the type of format that Excel automatically applied to each cell.

**Figure 2.7**
Testing Excel's default
formatting.

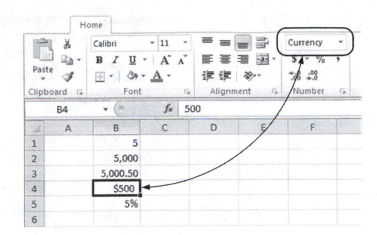

*Answers:*

- 5             General format
- 5,000         Number format (because of the comma)
- 5,000.50      Number format, two displayed decimal places
- $500          Currency format
- 5%            Percentage format

## PRACTICE!

Practice formatting numeric values.

Open a blank worksheet, and then follow these steps:

1. Enter 12023.45 into cells B1 through B4 as shown in Figure 2.8. By default, the cells are formatted in General format.

**Figure 2.8**
Values before applying
formatting options.

| | A | B | C | D |
|---|---|---|---|---|
| 1 | | 12023.45 | | |
| 2 | | 12023.45 | | |
| 3 | | 12023.45 | | |
| 4 | | 12023.45 | | |
| 5 | | | | |

First, format the value in cell B2 using a Number format with no displayed decimal places.

2. Right-click cell B2, and choose Format Cells . . . from the Quick Edit menu. The Format Cells dialog box will appear (Figure 2.9). Choose the **Number** tab and select "Number" from the list labeled **Category**.

3. Set the scroll box titled **Decimal places** to 0, and check the box labeled **Use 1000 Separator**. Note that this action rounds off the *displayed* value to zero decimal places, but the value *stored* in the cell is not actually changed. If you were to change the number of decimal places to 2, the fractional part of the number would be displayed.

Next, format cell B3 in Currency format.

**Figure 2.9**
Setting the number of displayed decimal places.

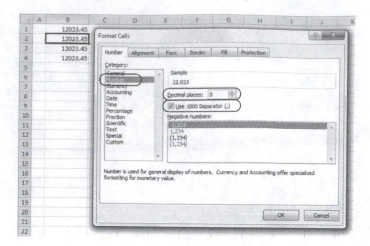

4. Right-click cell B3, and choose Format Cells . . . from the Quick Edit menu. The Format Cells dialog box will appear. Choose the Number tab and select Currency from the list.
5. Select 2 decimal places and select the $ symbol.
   Next, format cell B4 in *Scientific format*. This will display the number in scientific notation, base 10. For example, the number 256 is equal to $2.56 \times 10^2$, which is represented in Scientific format as $2.56E + 02$.
6. Right-click cell B4, and choose Format Cells . . . from the Quick Edit menu. The Format Cells dialog box will appear. Choose the Number tab and select Scientific from the list.
7. Select 6 decimal places.

Your resulting worksheet should resemble Figure 2.10.

**Figure 2.10**
Values after various formats have been applied.

|   | A | B | C | D |
|---|---|---|---|---|
| 1 |   | 12023.45 |   |   |
| 2 |   | 12,023 |   |   |
| 3 |   | $12,023.45 |   |   |
| 4 |   | 1.202345E+04 |   |   |
| 5 |   |   |   |   |

## 2.2.2 Entering Text Data

Text is commonly used for titles, headings, and labels. Such text data are entered into cells in the same ways that numeric values are entered. When Excel determines that a cell contains text data, it automatically applies the General format to the cell. (Notice that the General format is Excel's default for both numbers and text.) By default, cells containing text data are left-justified.

If you need to change the contents of a cell, simply double-click on the cell and you will be able to edit the text. Alternatively, select the cell and press the **F2** key to edit the text.

### 2.2.2.1 How Excel Determines That a Cell Contains Text

Excel determines that a cell contains text data by a process of elimination—a cell entry that is not a number, a date, a time, or a formula is treated as text. A cell that begins with A–Z, a–z, or a single quote ['] will always be treated as text. You can force a cell's contents to be treated as text by entering a single quote before the rest of the cell's contents. When the single quote is used, it will not be displayed but it will cause the cell's contents to be treated as text.

Examples:

- Entering 5.0 in a cell will be interpreted by Excel as numeric data. The General format will be applied, and the value will be displayed right-justified with no displayed decimal places.
- By including a single quote and entering '5.0 in a cell, Excel will interpret the entry as text data. The General format will be applied and the text value will be displayed left-justified as 5.0. (The single quote will not be displayed.)

Cells containing text data can also be formatted to change *font size*, color, etc. Applying formatting to change the appearance of displayed information will be presented later.

### 2.2.2.2 Using the Text Format

Another way to force a cell's contents to be treated as text data is to apply the *Text format* to the cell. To apply Text format to one or more cells, follow these steps:

1. Right-click on the cell (or selected cell range) to be formatted and choose **Format Cells . . .** from the Quick Edit menu. The Format Cells dialog box will appear.
2. Choose the **Number** tab on the Format Cells dialog box.
3. Select *Text* from the **Category** list.

**Note:** It is usually preferable to use the General format for text data.

---

### EXAMPLE 2.1

## HANDLING MISSING DATA

There are times when a complete data set is not available when you are developing your worksheet. By marking missing data as "not available" (or any other non-numeric text) you can create worksheets that automatically update as the missing data is added.

In this example, exam scores for a group of students have been entered on a worksheet (Figure 2.11). One of the students, Andy, was traveling with a team when the exam was given and will take a makeup exam when he gets back. We want the worksheet to automatically recalculate the exam average after his score is added.

**Figure 2.11**
Example of using text data to handle incomplete data sets.

| | A | B | C | D |
|---|---|---|---|---|
| 1 | | | | |
| 2 | | Name | Score | |
| 3 | | | | |
| 4 | | Ali | 85 | |
| 5 | | Andy | not available | |
| 6 | | Brittnay | 82 | |
| 7 | | Eduard | 80 | |
| 8 | | Erika | 88 | |
| 9 | | | | |
| 10 | | Average: | 83.75 | |
| 11 | | Count: | 4 | |
| 12 | | | | |

This example uses an Excel built-in function, called *AVERAGE*, to calculate the average of the values in a group of cells. One of the features of this function is that it ignores non-numeric data when calculating the average.

Andy's score is marked as "not available" when the scores are entered. Since "not available" is non-numeric, his score will be ignored when computing the average score.

The Count result illustrates that only four scores were included when calculating the average.

When Andy's score becomes available it is entered into cell C5, and the average exam score will be automatically recalculated, as shown in Figure 2.12.

**Figure 2.12**
The worksheet automatically updates when the missing data are added.

| | A | B | C | D |
|---|---|---|---|---|
| 1 | | | | |
| 2 | | Name | Score | |
| 3 | | | | |
| 4 | | Ali | 85 | |
| 5 | | Andy | 96 | |
| 6 | | Brittnay | 82 | |
| 7 | | Eduard | 80 | |
| 8 | | Erika | 88 | |
| 9 | | | | |
| 10 | | Average: | 86.2 | |
| 11 | | Count: | 5 | |
| 12 | | | | |

Notes:

- The worksheet will not update if the "not available" cells are formatted with the Text format. General format has been used on all cells in this example.
- There is nothing special about the term "not available." Any non-numeric text can be used.

### 2.2.3 Entering Date and Time Data

Excel stores dates and times internally as numbers. This allows you to perform arithmetic on dates and times. For example, you can subtract one date from another to determine the number of days between two events.

#### 2.2.3.1 How Excel Determines That a Cell Contains a Date or Time

Excel attempts to interpret entered data containing dashes (−) or slashes (/) as dates, and entered data containing a colon (:) as a time. Excel also tries to recognize standard abbreviations for months and days. For example, each of the "As entered" cell entries in Table 2.1 will be recognized as dates and/or times by Excel.

#### 2.2.3.2 Automatic Conversion to Date-Time Values

When Excel determines that an entered value is a date or a time, it automatically converts the entry to a number, called a ***date-time value***, and applies a ***Date format*** (if you enter a numeric date, such as 12/25/05) or a *Custom format* (if Excel interprets the date from text) to display the date and/or time. In a date-time value, the

**Table 2.1  How Excel Interprets Entered Dates and Times**

| As Entered . . . | Recognized as . . . | Converted to . . . | Displayed as . . . |
|---|---|---|---|
| December 25, 2005 | Date | 38711.00000 | 25-Dec-05 |
| 12/25/05 | Date | 38711.00000 | 25-Dec-05 |
| Mar 3, 2007 | Date | 39144.00000 | 3-Mar-07 |
| 8:15 | Time (a.m.) | 0.34375 | 8:15 |
| 8:15 pm | Time | 0.84375 | 20:15 |
| 8:15 am, Mar 3, 2007 | Date and Time | 39144.34375 | 3/3/2007 8:15 |

decimal places are used to indicate time as a fraction of 24 hours. For example, noon has a time value of 0.5, since it is halfway through the 24-hour day.

The digits to the left of the decimal point are used to indicate the number of days since the starting date that was used to compute date-time values. On PCs, the default start date is January 1, 1900. On Macintosh computers, the default start date is January 1, 1904. You can determine which date system is in use on your computer by entering a value of 1 in a cell and then changing the cell's format to Date format. The default start date will be shown in the cell as shown in Figure 2.13 (lower panel).

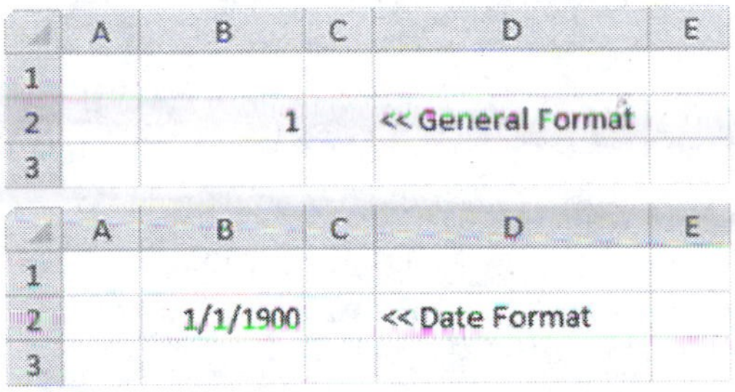

**Figure 2.13**
Checking the start date.

Excel allows you to choose either start date from the Options menu. Use the **Microsoft Office button**, then click the **Excel Options button** and select the **Advanced group**. Select or clear the **Use 1904 Date System** box in the **When Calculating this Workbook** section. [Excel 2003: Use **Tools → Options**, and then select the **Calculation tab**. Check or uncheck the box labeled **1904 date system**.]

The automatic conversion of dates and times to date-time values is one of the few instances in which Excel actually changes the contents of a cell, not just the way the contents are displayed. This can be frustrating if Excel interprets the value you are entering into a cell as a date or time and automatically changes the value you entered. This would only happen if you were entering a fraction, such as 4/5. The presence of the slash would cause Excel to interpret the entry as a date (April 5), convert the value to a date-time value, and display April 5 of the current year. To enter a fraction, first set the cell's format to the *Fraction format*, then and enter the value.

### 2.2.3.3 Date and Time Display Formats

When Excel detects that a date or time has been entered into a cell, the date-time value that is stored in the cell is displayed, using the format specified by the **Regional and Language Options** (**Control Panel → Regional and Language Options**) in Microsoft Windows, not Excel. This is implemented in Excel by using a Custom format for each date and time.

If you want to apply a format to a cell containing a date or time, you will probably use the Date format or *Time format* rather than creating a custom format.

A Date format can be applied to a cell or range of cells, as follows:

1. Select the cell or range of cells to be formatted.
2. Right-click the selected cell(s) and select **Format Cells . . .** from the Quick Edit menu. The Format Cells dialog box will appear.
3. Choose the **Number** tab.
4. Select *Date* from the **Category** list.
5. Choose the display format you would like Excel to use from the **Type** list.

To apply a Time format to a range of cells, follow these steps:

1. Select the cell or range of cells to be formatted.
2. Right-click the selected cell(s) and select **Format Cells . . .** from the Quick Edit menu. The Format Cells dialog box will appear.
3. Choose the **Number** tab.
4. Select *Time* from the **Category** list.
5. Choose the display format you would like Excel to use from the **Type** list.

## PRACTICE!

Practice entering dates and times in ways that Excel can recognize.

Enter a date in several standard ways to see how many Excel can recognize and convert to date-time values. For example, try December 25, 2025 in the following ways:

- 12/25/2025
- 12/25/25
- 12-25-2025
- Dec-25-25
- December 25, 2025
- 25 December 2025

The results are shown in Figure 2.14.

**Figure 2.14**
Testing various ways
of entering dates.

| | As Entered | As Displayed | Comment |
|---|---|---|---|
| | 12/25/2025 | 12/25/2025 | Recognized as a date, displayed with a Date format |
| | 12/25/25 | 12/25/2025 | " |
| | 12-25-2025 | 12/25/2025 | " |
| | Dec-25-25 | Dec-25-25 | Not recognized as a date |
| | December 25, 2025 | 25-Dec-25 | Recognized as a date, displayed with a Custom format |
| | 25 December 2025 | 25-Dec-25 | " |

## 2.3 USING THE FILL HANDLE

Data entry can be tedious. The use of the *Fill Handle* allows you to quickly copy a cell into a row or column of cells. Fill Handles can also be used to create a series of numbers in a row or column.

The Fill Handle appears as a small black square in the bottom-right corner of a selected region. An example of a Fill Handle is shown in Figure 2.15.

**Figure 2.15**
The Fill Handle is located at the bottom-right corner of a selected region.

Fill Handle

### 2.3.1 Using the Fill Handle with the Left Mouse Button to Copy Cells

When the cursor is placed over a Fill Handle, its shape will change to a black cross. Click and hold the left mouse button while dragging the fill handle to the right four or five cells. When the mouse button is released, the value in the original cell will be copied into the new cells, as shown in Figure 2.16.

**Figure 2.16**
Copying cells using the Fill Handle.

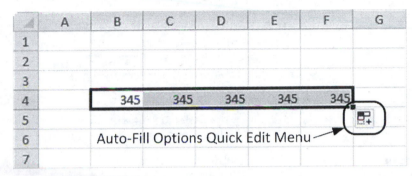

Auto-Fill Options Quick Edit Menu

### 2.3.2 Using the Fill Handle with the Left Mouse Button—More Copy Options

After copying the cells, a small button will appear near the Fill Handle; this is called the *Auto-Fill Options Quick Edit Menu* (Figure 2.16). If you click on the Quick Edit Menu button, some options for filling the new cells will be displayed. This is illustrated in Figure 2.17.

**Figure 2.17**
Using the Auto-Fill Options Menu to create a series of values.

Using the Auto-Fill Options Menu, you can select from the following items:

* Copy Cells—simply copies the value (and formatting) of the original cell into the new cells.
* Fill Series—increments the value in the original cell to create a series in the new cells.

- Fill Formatting Only—copies only the format of the original cell to the new cells; the values of the new cells are left unchanged.
- Fill Without Formatting—copies the value of the original cell into the new cells, but does not change the formatting of the new cells.

### 2.3.3 Creating a Linear Series with a Non-Unity Increment

As shown in Figure 2.17, it is easy to use the Fill Handle to create a series in which the values are incremented by one. However, there are many instances where an increment value other than one is needed. To specify a series increment, simply enter the first two values of the series before selecting the cells to be copied. This is illustrated in Figure 2.18.

**Figure 2.18**

Preparing to create a series in which values are incremented by 5.

|   | A | B | C | D | E | F | G |
|---|---|---|---|---|---|---|---|
| 1 |   |   |   |   |   |   |   |
| 2 |   | 345 | 350 |   |   |   |   |
| 3 |   |   |   |   |   |   |   |

Once the first two values of the series are selected, drag the Fill Handle with the left mouse button to complete the series (Figure 2.19).

**Figure 2.19**

Completing the series using the Fill Handle.

|   | A | B | C | D | E | F | G |
|---|---|---|---|---|---|---|---|
| 1 |   |   |   |   |   |   |   |
| 2 |   | 345 | 350 | 355 | 360 | 365 |   |
| 3 |   |   |   |   |   |   |   |
| 4 |   |   |   |   |   |   |   |

### 2.3.4 Using the Fill Handle with the Right Mouse Button for Additional Options

If you drag the Fill Handle with the right mouse button, a menu of fill options will automatically pop up when the mouse button is released. The first four menu options, shown in Figure 2.20, are the same as the ones available through the Auto-Fill Options Menu. However, additional fill options are also provided.

**Figure 2.20**

Fill options available with the right mouse button.

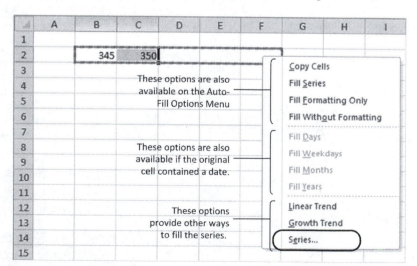

The new fill options include a Linear Trend option that creates a linear series of values (with an increment of 5) and a Growth Trend option that creates an exponential growth series (fit to the two original values). The last menu item opens the Series dialog box, as shown in Figure 2.21.

**Figure 2.21**
The Series dialog box.

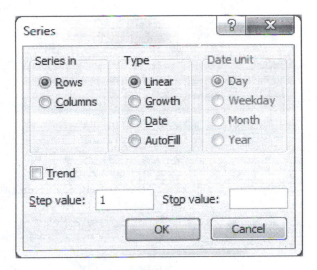

This dialog provides another way of specifying an increment or step value for a series, as well as allowing both linear and exponential growth series to be created.

## PRACTICE!

Practice using Fill Handles and creating Fill Series by performing the steps that follow. Note that sometimes you will be using the left mouse button and sometimes you will be using the right mouse button.

First, practice copying the contents of a cell:

1. Open a new worksheet.
2. Type the value 1.5 into cell A1. Grab the Fill Handle for cell A1 with the left mouse button and drag it over cells B1:G1.
3. When you release the mouse button, the value 1.5 will have been copied to cells B1:G1, as shown in Figure 2.22.

| | A | B | C | D | E | F | G | H |
|---|---|---|---|---|---|---|---|---|
| 1 | 1.5 | 1.5 | 1.5 | 1.5 | 1.5 | 1.5 | 1.5 | |
| 2 | | | | | | | | |

**Figure 2.22**
The result after Step 3.

Now try creating a linear series:

4. Type the values 1.5 and 1.7 into cells A3 and B3, respectively.
5. Select the region A3:B3 and grab the Fill Handle with the left mouse button.
6. Drag the fill handle over the cells C3:G3. When you release the mouse button, a linear series will have been created.

Note (see Figure 2.23) that Excel has created the linear Fill Series by using the difference between the first two cells (A3 and B3) as the increment. You can also control the increment value using the Series dialog, as illustrated with the following steps:

| | A | B | C | D | E | F | G | H |
|---|---|---|---|---|---|---|---|---|
| 3 | 1.5 | 1.7 | 1.9 | 2.1 | 2.3 | 2.5 | 2.7 | |
| 4 | | | | | | | | |

**Figure 2.23**
The linear series created with the Fill Handle.

7. Type the value 1.5 into cell A5.
8. Grab the Fill Handle for cell A5 with the right mouse button and drag it over cells B5:G5.
9. When you release the mouse button, the Fill Series drop-down menu will appear.
10. Select *Series* from the drop-down menu. The Series dialog box will appear.
11. Check *Rows* in the box labeled **Series in**.
12. Check *Linear* in the box labeled **Type**.
13. Type **0.4** in the box labeled **Step value**.
14. Click **OK**.

The cells will be filled with a linear series in increments of 0.4, as shown in Figure 2.24.

| | A | B | C | D | E | F | G | H |
|---|---|---|---|---|---|---|---|---|
| 5 | 1.5 | 1.9 | 2.3 | 2.7 | 3.1 | 3.5 | 3.9 | |
| 6 | | | | | | | | |

**Figure 2.24**
The linear series with an increment of 0.4 was created using the Series dialog box.

## 2.4 FORMATTING FOR APPEARANCE

Several formatting options are available to change the appearance of information in your worksheet. These include fonts, colors, borders, and text alignment. Careful application of these formatting options can help you create professional-looking worksheets, and help focus attention on the most important results.

Some appearance changes can be applied to selected cell ranges, while additional characteristics, such as *column width* or row height, apply only to selected columns or rows. Because of this, formatting for cells, columns and rows, and the entire worksheet will be presented separately.

### 2.4.1 Changing the Appearance of Cells

In previous sections, format options were selected from the Format Cells dialog box. This dialog box can also be used to change the appearance of cells; however, many of the most commonly used format changes are also available on the Ribbon's **Home tab** [Excel 2003: **Format bar**]. When possible, use of the Ribbon will be emphasized here.

**Figure 2.25**
The Excel Ribbon's Home tab.

The Ribbon can be customized, but typically the Home tab looks something like Figure 2.25. The amount of information displayed on the Ribbon changes as the size of the Excel window changes, so the Ribbon may look a little different on your computer.

### 2.4.1.1 Home/Font Group: Changing Text Attributes

Using the Ribbon, you can quickly change the appearance of your worksheet. Near the left side of the Home tab (Figure 2.25) is the *Font Group* (Figure 2.26) which contains drop-down lists you can use to select a **font style** and font size. The selected font style or size is applied to the cells that are selected when the font is changed.

**Figure 2.26**
The Font Group on the Ribbon's Home tab.

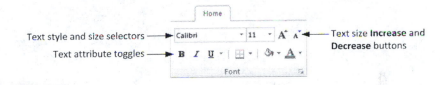

Below the drop-down lists are three toggle buttons that allow you to apply and remove bold, italic, and underline attributes from the text in selected cells.

**Note:** The term *toggle button* is used to indicate that you click the button once to activate the attribute then click the same button again to deactivate it. For example, to show the displayed value in a cell in a bold font, first select the cell, then click the Bold toggle button on the Ribbon (Home tab, Font group). This activates the bold attribute for the cell. If you change your mind, simply select the cell and click the Bold button again to deactivate the bold attribute for the cell.

### 2.4.1.2 Home/Font Group: Changing Background and Text Colors

The Font Group on the Home tab also allows you to change cell background colors and text colors using the buttons indicated in Figure 2.27.

**Figure 2.27**
Cell fill color and text color buttons on the Font Group.

If you click either of the color selection buttons, the currently selected color (shown on the button) is applied to the selected cells. To use a different color, click the down arrow on the right edge of the buttons; this causes a color palette to be displayed, as shown in Figure 2.28.

**Figure 2.28**

Selecting a new color from the color palette.

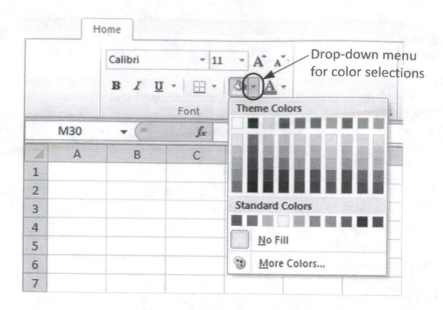

### 2.4.1.3 Home/Font Group: Changing Cell Borders

The Format Toolbar also allows you to change *cell borders* using the button indicated in Figure 2.29.

**Figure 2.29**

Changing cell borders.

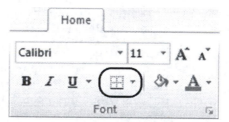

Clicking the cell border button applies the border indicated on the button (lower edge, in Figure 2.29) to the currently selected cell or cell range. To change the location or style of border, click the down arrow on the border button to see a palette of border options, as shown in Figure 2.30.

**Note:** To get rid of an existing border, use the **No Border** option.

There are some limitations when using the Border Options Palette from the Format bar:

- The borders will always be drawn in the current border color (black, by default).
- There are border options such as "only interior borders" that are not available on the palette.

You have two options for greater control over cell borders. You can:

1. Activate **Draw Borders** from the Border Options Palette.
2. Click the **More Borders . . .** button on the Border Options Palette to open the Format Cells dialog, Border panel, shown in Figure 2.31.

Both of these options give you control over color, style, and placement of borders on selected cells and selected cell ranges.

**Figure 2.30**
Border Options Palette.

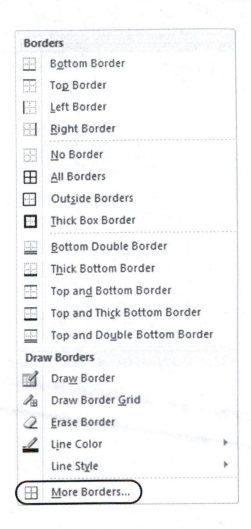

**Figure 2.31**
Changing the border
characteristics using the
Format Cells dialog, Border
panel.

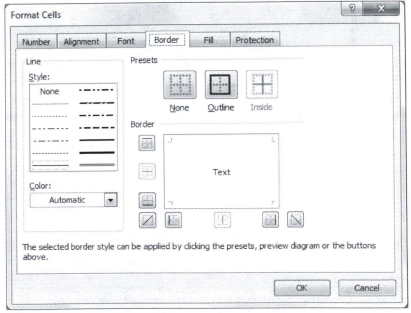

### 2.4.1.4 Home/Alignment Group

The toggle buttons in the **Alignment group** (see Figure 2.32) allow you to specify how the cell contents are displayed within the cells.

**Figure 2.32**

The Alignment group on the Ribbon's Home tab.

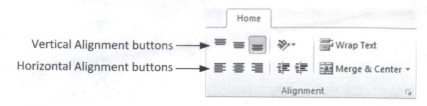

You can set both *vertical* and *horizontal alignments*.

**Vertical Alignment**

- Top align
- Middle Align
- Bottom Align (default)

**Horizontal Alignment**

- Left justify (default for text)
- Center justify
- Right justify (default for numbers)

The effect of various alignments is illustrated in Figure 2.33.

**Figure 2.33**

Alignment options.

| ▲ | A | B | C | D | E | F |
|---|---|---|---|---|---|---|
| 1 | | | | | | |
| 2 | | | Left Align | Center Align | Right Align | |
| 3 | | Top Align | text | text | text | |
| 4 | | Middle Align | text | text | text | |
| 5 | | Bottom Align | text | text | text | |
| 6 | | | | | | |

The Alignment group on the Ribbon's Home tab also provides easy ways to wrap text within cells, and center a heading in a number of merged cells (Figure 2.34).

**Figure 2.34**

Buttons used to wrap text in cells and center a heading in multiple cells.

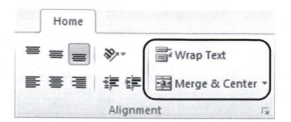

Both of these options are commonly used with table headings. For example, Figure 2.35 shows a loan table that can be used to determine how much you owe after each car payment, but the column headings in row 9 are pretty much unreadable because they are too long for the cells.

**Figure 2.35**
Loan table before improving the headings.

| | A | B | C | D | E | F |
|---|---|---|---|---|---|---|
| 1 | Car Loan Table | | | | | |
| 2 | | | | | | |
| 3 | Amount Borrowed: | | $ 15,000.00 | | | |
| 4 | | APR: | 8% | | | |
| 5 | | Term: | 2 | years | | |
| 6 | Monthly Payment: | | $678.41 | | | |
| 7 | | | | | | |
| 8 | | | Amounts in Dollars | | | |
| 9 | Payment I | Loan Amount | Paid on Inter | Paid on Princ | Loan Amount After Pay | |
| 10 | 1 | 15000.00 | 100.00 | 578.41 | 14421.59 | |
| 11 | 2 | 14421.59 | 96.14 | 582.27 | 13839.33 | |
| 12 | 3 | 13839.33 | 92.26 | 586.15 | 13253.18 | |
| 13 | 4 | 13253.18 | 88.35 | 590.05 | 12663.12 | |
| 14 | 5 | 12663.12 | 84.42 | 593.99 | 12069.13 | |
| 30 | 21 | 2669.01 | 17.79 | 660.62 | 2008.39 | |
| 31 | 22 | 2008.39 | 13.39 | 665.02 | 1343.37 | |
| 32 | 23 | 1343.37 | 8.96 | 669.45 | 673.92 | |
| 33 | 24 | 673.92 | 4.49 | 673.92 | 0.00 | |
| 34 | | | | | | |

The improved table, shown in Figure 2.36, is easier to read because the heading text in row 9 has been wrapped. Notice that the **Amounts in Dollars** label has been *merged and centered* over the values in columns B through E.

**Figure 2.36**
Improved loan table with readable headings.

| | A | B | C | D | E | F |
|---|---|---|---|---|---|---|
| 1 | Car Loan Table | | | | | |
| 2 | | | | | | |
| 3 | Amount Borrowed: | | $ 15,000.00 | | | |
| 4 | | APR: | 8% | | | |
| 5 | | Term: | 2 | years | | |
| 6 | Monthly Payment: | | $678.41 | | | |
| 7 | | | | | | |
| 8 | | | Amounts in Dollars | | | |
| 9 | Payment Number | Loan Amount Before Payment | Paid on Interest | Paid on Principal | Loan Amount After Payment | |
| 10 | 1 | 15000.00 | 100.00 | 578.41 | 14421.59 | |
| 11 | 2 | 14421.59 | 96.14 | 582.27 | 13839.33 | |
| 12 | 3 | 13839.33 | 92.26 | 586.15 | 13253.18 | |
| 13 | 4 | 13253.18 | 88.35 | 590.05 | 12663.12 | |
| 14 | 5 | 12663.12 | 84.42 | 593.99 | 12069.13 | |
| 30 | 21 | 2669.01 | 17.79 | 660.62 | 2008.39 | |
| 31 | 22 | 2008.39 | 13.39 | 665.02 | 1343.37 | |
| 32 | 23 | 1343.37 | 8.96 | 669.45 | 673.92 | |
| 33 | 24 | 673.92 | 4.49 | 673.92 | 0.00 | |
| 34 | | | | | | |

### 2.4.1.5 Home/Number Group

The Number group on the Ribbon's Home tab provides access to number formats (General, Numeric, Currency, etc.) as well as some useful toggle buttons, as shown in Figure 2.37.

**Figure 2.37**

The Number group on the Ribbon's Home tab.

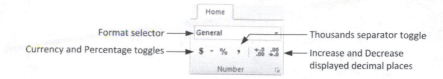

- The **currency toggle button** allows you to display a value with a dollar sign (or other currency symbol) and two decimal places.
- The **percentage toggle** displays a value as a percentage by moving the decimal point two places to the right and showing the percent symbol. For example, the value 0.75 would be displayed as 75% if the percentage format has been toggled on.
- Clicking the **thousands separator button** causes the value to be displayed in Accounting format with *thousands separators* (commas) and two decimal places. For example, the value 12023.4 would be displayed as 12,023.40 if the thousands separator format has been toggled on.
- **Increase the number of displayed decimal places** toggle button.
- **Decrease the number of displayed decimal places** toggle button.

The two buttons that increase and decrease the number of displayed decimal places can be very useful when you want all of the values in a list to show the same number of decimal places:

1. Select the entire set of values.
2. Increase or decrease the number of displayed decimal places on the first item in the list to display the desired number of decimal places.

All of the selected values will end up showing the same number of decimal places.

For example, when you create a column of values that are uniformly increasing by 0.2, the whole numbers will, by default, show one less decimal place. This is illustrated in Figure 2.38.

**Figure 2.38**

The selected list.

| ⁄ | A | B | C | D | E |
|---|---|---|---|---|---|
| 1 | | | | | |
| 2 | | 1 | | | |
| 3 | | 1.2 | | | |
| 4 | | 1.4 | | | |
| 5 | | 1.6 | | | |
| 6 | | 1.8 | | | |
| 7 | | 2 | | | |
| 8 | | 2.2 | | | |
| 9 | | 2.4 | | | |
| 10 | | | | | |

The *active cell* is a different color from the rest of the cells in the selected range.

To make the entire list show two decimal places, you do the following:

1. Select the entire list, cells B3:B10.
2. Click on the **Increase Decimal** button twice.

The number of displayed decimal places of the currently active cell (cell B3 in Figure 2.39) will be increased by two (because the button was clicked twice), and the entire list will be displayed with two decimal places.

**Figure 2.39**
The result—all values displayed with two decimal places.

| | A | B | C | D | E |
|---|---|---|---|---|---|
| 1 | | | | | |
| 2 | | 1.00 | | | |
| 3 | | 1.20 | | | |
| 4 | | 1.40 | | | |
| 5 | | 1.60 | | | |
| 6 | | 1.80 | | | |
| 7 | | 2.00 | | | |
| 8 | | 2.20 | | | |
| 9 | | 2.40 | | | |
| 10 | | | | | |

## EXAMPLE 2.2

### FORMATTING A WORKSHEET

In the following example (Figure 2.40), the area of a triangle is being calculated from entered values for the length of the base, b, and the height, h, according to the following formula:

$$Area = \frac{1}{2}bh$$

**Figure 2.40**
Formatting example.

| | A | B | C | D | E | F | G | H |
|---|---|---|---|---|---|---|---|---|
| 1 | Area of a Triangle | | | | | | | |
| 2 | | | | | | | | |
| 3 | | b: | 1 | cm | | | | |
| 4 | | h: | 3 | cm | | | | |
| 5 | | | | | | | | |
| 6 | | Area: | 1.5 | cm2 | | | | |
| 7 | | | | | | | | |

To improve the appearance and readability of the worksheet, we will make the following formatting changes:

1. Make the title in cell A1 stand out by increasing the font size and adding the bold attribute.
2. Move the labels in column B closer to the values by right-justifying the labels.
3. Place a border around the calculated result in cell C6.
4. Superscript the 2 in cell D6.

First, to format the title in cell A1:

1. Select cell A1.
2. Select 12-point font from the font size drop-down list in the Font Group on the Ribbon's Home tab (Figure 2.41).
3. Click the Bold button.

**Figure 2.41**
Formatting the title in cell A1.

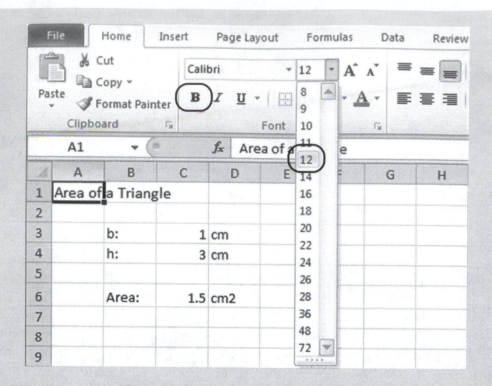

To right justify the labels in column B:

4. Select the labels in cells B3:B6 as shown in Figure 2.42.
5. Click the Align Right button to right justify the labels.

**Figure 2.42**
Right aligning the labels in column B.

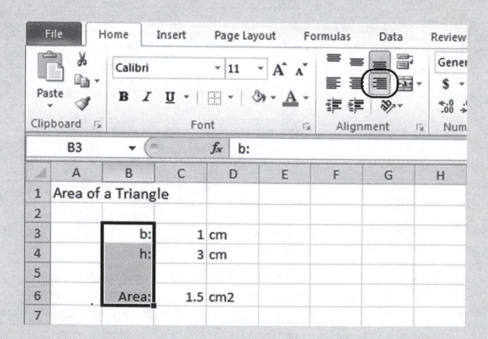

To put a border around cells C6 and C7 to highlight the result:

6. Select the cell containing the calculated result, cell C6, and the cell containing the units, D6, as shown in Figure 2.43.
7. Click the drop-down arrow on the right side of the border button in the Font group.
8. Select a full outside border from the border options menu as shown in Figure 2.43.

**Figure 2.43**
Adding a border around the calculated result.

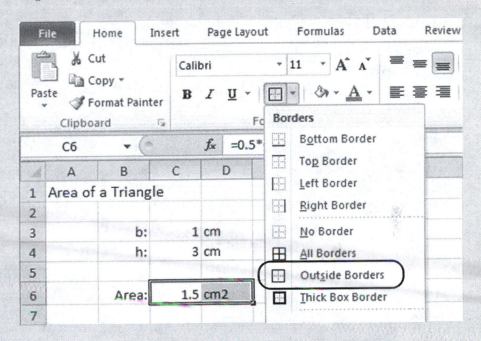

The result of these format changes is shown in Figure 2.44.

**Figure 2.44**
The result of the formatting changes.

| ⏴ | A | B | C | D | E | F | G | H |
|---|---|---|---|---|---|---|---|---|
| 1 | Area of a Triangle | | | | | | | |
| 2 | | | | | | | | |
| 3 | | b: | 1 | cm | | | | |
| 4 | | h: | 3 | cm | | | | |
| 5 | | | | | | | | |
| 6 | | Area: | 1.5 | cm2 | | | | |
| 7 | | | | | | | | |

Still need to superscript the 2 in $cm^2$

The last desired format change, superscripting the 2 in cell D6, cannot be performed using buttons on the Ribbon. Instead, the change must be made from the Format Cells dialog box. You can open the Format Cells dialog box using the expansion button (indicated in Figure 2.45) on the Font group.

*(continued)*

**Figure 2.45**
Accessing the Format Cells dialog box from the Font group.

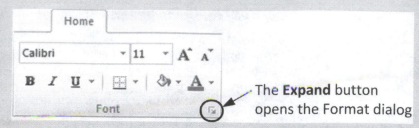

The **Expand** button opens the Format dialog

Superscripting the 2 in $cm^2$ in cell D6.

1. Double-click on cell D6 and select just the 2 in cm2.
2. Click on the expansion button in the Font group to open the Format Cells dialog, shown in Figure 2.46.
3. Check the **Superscript** box in the **Effects** list, as illustrated in Figure 2.46.

**Figure 2.46**
The Format Cells dialog box.

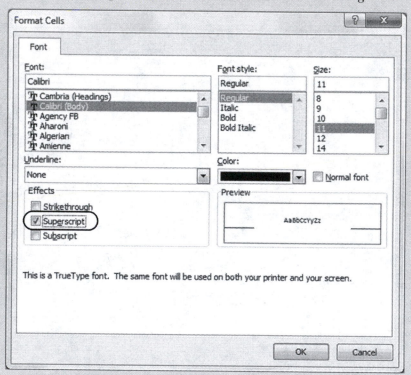

The final result is shown in Figure 2.47.

**Figure 2.47**
The final result.

|  | A | B | C | D | E | F | G | H |
|---|---|---|---|---|---|---|---|---|
| 1 | Area of a Triangle | | | | | | | |
| 2 | | | | | | | | |
| 3 | | b: | 1 | cm | | | | |
| 4 | | h: | 3 | cm | | | | |
| 5 | | | | | | | | |
| 6 | | Area: | 1.5 | cm$^2$ | | | | |
| 7 | | | | | | | | |

While the formatting changes applied to this worksheet are purely superficial, they do make the worksheet easier to read and more professional in appearance, and they cause the intent and result of the calculation to be more readily apparent.

## ENGINEERING ECONOMICS

Engineering economics involves the study of interest, cash flow patterns, depreciation, and inflation, and techniques for maximizing net value. This is an important area of study for all engineers, since engineers frequently serve as managers or executive officers of corporations.

The next example demonstrates how John can make a choice between two investment alternatives. One option is to place $10,000 in a very secure investment that will give him 6% annual growth. The other option is to put the $10,000 in the stock market, using a fund that has historically shown 11% growth. John knows that there is no guarantee that the stock fund will continue to return a rate of 11%, and he could even lose money. How should he decide which is the best investment for his money?

Well, the decision will ultimately be up to John and based on the level of risk that he is willing to accept. Excel, however, can help him to forecast the growth differential for the two options, and he can use this information to help him decide if it is worth it to take a chance on the stock market. The results of his calculations are shown in Figure 2.48.

**Figure 2.48**

The results of John's analysis.

| Accumulated Capital | | |
|---|---|---|
| Age | 6% Growth | 11% Growth |
| 18 | $ 10,000 | $ 10,000 |
| 28 | $ 17,908 | $ 28,394 |
| 38 | $ 32,071 | $ 80,623 |
| 48 | $ 57,435 | $ 228,923 |
| 58 | $ 102,857 | $ 650,009 |

John's analysis shows that, if the stock market fund does return 11% annual growth, in 40 years he will end up with over six times more money than if he puts the $10,000 in the low risk, 6% fund. Now he has a decision to make!

**How Did He Do That?**

The future value of an amount invested in an interest-bearing fund can be calculated with the following equation:

$$F = P[1 + i]^N$$

Here, $F$ is the future value, $P$ is the amount initially deposited, $i$ is the annual interest rate, and $N$ is the number of years that the funds remain invested.

John built this equation into his worksheet and had Excel calculate future values for two different interest rates and four different values of N.

**How Did He Make His Table Look So Good?**

Initially, John's calculated results looked like Figure 2.49, but he made some formatting changes to improve the appearance. For practice, try making these format changes yourself. Begin by typing in the data shown in Figure 2.49, and then complete the steps to make your table look like the table in Figure 2.48.

**Figure 2.49**
John's calculated results before formatting.

| | A | B | C | D |
|---|---|---|---|---|
| 1 | Accumulated Capital | | | |
| 2 | Age | 6% Growth | 11% Growth | |
| 3 | 18 | 10000 | 10000 | |
| 4 | 28 | 17908 | 28394 | |
| 5 | 38 | 32071 | 80623 | |
| 6 | 48 | 57435 | 228923 | |
| 7 | 58 | 102857 | 650009 | |
| 8 | | | | |

## Add Dollar Signs to the Values

To add dollar signs to the values in columns B and C, do the following:

1. Select the currency values in cell range B3:C7.
2. Choose **Accounting format** from the Ribbon [**Home tab → Number group → Number format drop-down list**]. This will add the currency symbols (dollar signs) and display the values with two decimal places.
3. While cell range B3:C7 is still selected, click the **Decrease Decimal** button [**Home tab → Number group**] twice to remove the displayed decimal places.

At this point the worksheet looks like Figure 2.50.

**Figure 2.50**
Results table after adding dollar signs.

| | A | B | C | D |
|---|---|---|---|---|
| 1 | Accumulated Capital | | | |
| 2 | Age | 6% Growth | 11% Growth | |
| 3 | 18 | $ 10,000 | $ 10,000 | |
| 4 | 28 | $ 17,908 | $ 28,394 | |
| 5 | 38 | $ 32,071 | $ 80,623 | |
| 6 | 48 | $ 57,435 | $ 228,923 | |
| 7 | 58 | $ 102,857 | $ 650,009 | |
| 8 | | | | |

## Center the Title over the Entire Table

To center the title over the table, we will merge cells A1 through C1, and then center the new merged cell by following these steps:

1. Select the region A1:C1.
2. Choose the **Merge & Center button** on the Ribbon [**Home tab → Alignment group → Merge & Center button**].

## Make the Column Headings Boldface

To make column headings in rows 1 and 2 boldface, do the following:

1. Select the column headings in rows 1 and 2 (cell range A1:C2).
2. Choose the **Bold** button on the Ribbon [**Home tab → Font group → Bold button**] (or, press **Ctrl + B**).

### Adjust the Width of the Columns

To *AutoFit* the column widths to the cell contents:

1. Select columns A, B, and C.
2. Position the mouse over the right edge of one of the columns (the mouse pointer will change to a vertical line with double-headed arrows).
3. Double-click the mouse.

The result is shown in Figure 2.51. Notice that column A was adjusted to fit the heading "Age" and not "Accumulated Capital." Merged cells are ignored when auto-fitting the column widths to the cell contents.

**Figure 2.51**

Results table after adjusting column widths.

| | A | B | C | D |
|---|---|---|---|---|
| 1 | | Accumulated Capital | | |
| 2 | Age | 6% Growth | 11% Growth | |
| 3 | 18 | $ 10,000 | $ 10,000 | |
| 4 | 28 | $ 17,908 | $ 28,394 | |
| 5 | 38 | $ 32,071 | $ 80,623 | |
| 6 | 48 | $ 57,435 | $ 228,923 | |
| 7 | 58 | $ 102,857 | $ 650,009 | |
| 8 | | | | |

### Center the Values

To center the values in the each column, follow these steps:

1. Select the cell range containing the column headings and the values A2:C7.
2. Click on the Center Alignment button on the Ribbon [**Home tab → Alignment group → Center Alignment button**].

### Add Borders Around Each Cell

To add borders around each cell, follow these steps:

1. Select the entire data set, cell range A1:C7.
2. Use the drop-down menu on the Borders button [**Home tab → Font group → Borders drop-down menu**].
3. Choose **All Borders** from the **Borders** menu.

### Change the Cell Fill Color and Text Colors of the Title

Follow these steps to change the background (fill) and text colors of the title:

1. Select the title (the merged cells A1:C1).
2. Use the drop-down menu on the Fill Color button [**Home tab → Font group → Fill Color drop-down menu**].
3. Choose a dark color for the cell background from the color palette. (Gray was selected in this example.)
4. Use the drop-down menu on the Font Color button [**Home tab → Font group → Font Color drop-down menu**].
5. Choose a light color for the title text. (White was selected in this example.)

Your completed table should now look like Figure 2.52.

**Figure 2.52**
The formatted table.

|  | A | B | C | D |
|---|---|---|---|---|
| 1 | Accumulated Capital | | | |
| 2 | Age | 6% Growth | 11% Growth | |
| 3 | 18 | $ 10,000 | $ 10,000 | |
| 4 | 28 | $ 17,908 | $ 28,394 | |
| 5 | 38 | $ 32,071 | $ 80,623 | |
| 6 | 48 | $ 57,435 | $ 228,923 | |
| 7 | 58 | $ 102,857 | $ 650,009 | |
| 8 | | | | |

## 2.4.2 Changing the Appearance of Columns and Rows

The primary formatting option that applies to an entire column is the column width. Similarly, you can set the height of entire rows. These values can be set in several ways:

- You can type the desired column width or row height directly into a dialog box.
- You can drag the edge of the column and row headings with the mouse to change their sizes.
- You can ask Excel to AutoFit the column width or row height to the contents of the column or row.

### 2.4.2.1 Using a Dialog Box to Set the Row Height
To change the height of an entire row of cells:

1. Select the row or rows that you want to modify.
2. Use Ribbon options **Home tab → Cells group → [Format] drop-down menu** and then click the **Row Height . . .** button, or right-click on the row header and select **Row Height . . .** from the pop-up menu. The latter approach is illustrated in Figure 2.53.

**Figure 2.53**
Accessing the Row Height dialog from the Cells group.

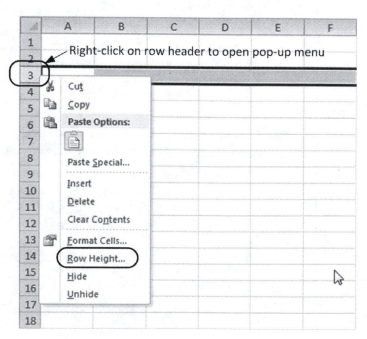

The *Row Height* dialog box (Figure 2.54) will open. Enter the desired new row height in the **Row height** field. (The default row height is 15.)

**Figure 2.54**

The Row Height dialog box.

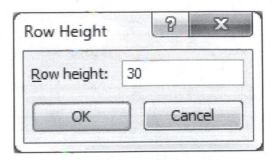

### 2.4.2.2 Using the Mouse to Change Column Widths

Column widths are easy to change with the mouse:

1. Select the columns that you want to modify by clicking on the column headings. In Figure 2.55 columns B through D have been selected.
2. Position the mouse at the right edge of any selected column heading. The mouse cursor will change to a vertical bar with left and right arrows when positioned over the edge of the column (indicated in Figure 2.55).

**Figure 2.55**

The mouse cursor when positioned over the edge of a column heading.

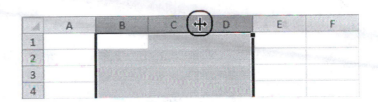

3. To adjust the column widths, hold the left mouse button down and drag the vertical bar cursor to the desired column width (Figure 2.56). A Tool Tip will be displayed showing the actual column width.

**Figure 2.56**

As the mouse cursor is dragged, only one column width is changed.

When you release the mouse button, the width of all selected columns will be changed (Figure 2.57).

**Figure 2.57**

When you release the mouse button, all selected columns are widened.

### 2.4.2.3 Automatically Fitting Column Widths to Column Contents

Excel will automatically adjust the width of selected columns (or the height of rows) to fit the contents of those columns (or rows). For example, in the worksheet shown in Figure 2.58, cell B2 contains a small number, and cell C2 contains a long text string.

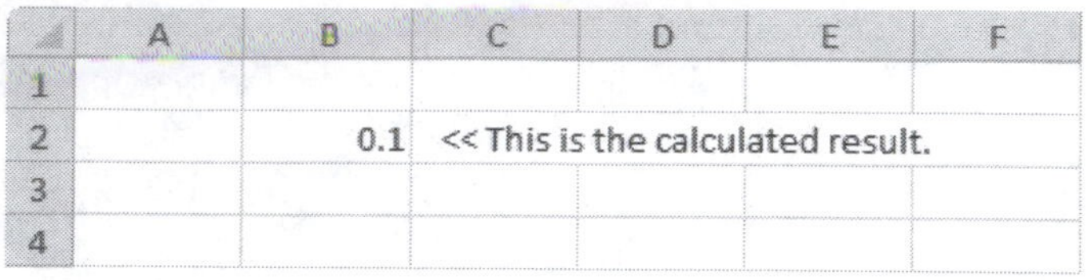

**Figure 2.58**
Before adjusting the columns widths.

| ▲ | A | B | C | D | E | F |
|---|---|---|---|---|---|---|
| 1 | | | | | | |
| 2 | | 0.1 | << This is the calculated result. | | | |
| 3 | | | | | | |
| 4 | | | | | | |

To automatically adjust the width of columns to the columns' contents:

1. Select the column(s) to be adjusted.
2. Position the cursor over the right edge of any selected column's heading. The mouse cursor changes to a crossbar with left and right arrows (Figure 2.59) when it is positioned over the edge of the column heading.

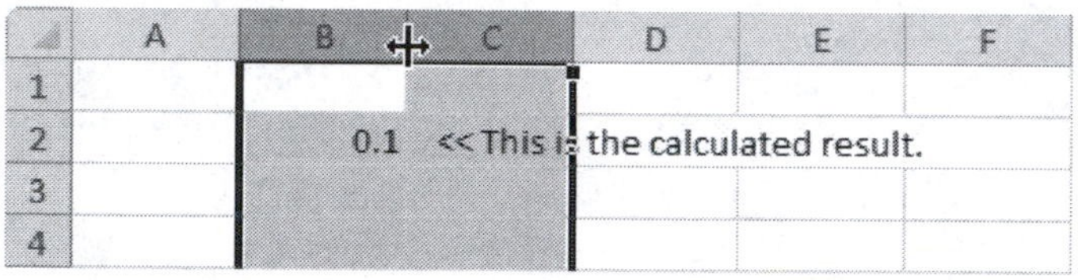

**Figure 2.59**
Preparing to change the column widths.

| ▲ | A | B | C | D | E | F |
|---|---|---|---|---|---|---|
| 1 | | | | | | |
| 2 | | 0.1 | << This is the calculated result. | | | |
| 3 | | | | | | |
| 4 | | | | | | |

3. Double-click the mouse's left button to adjust the column width to the column contents.

The result is shown in Figure 2.60.

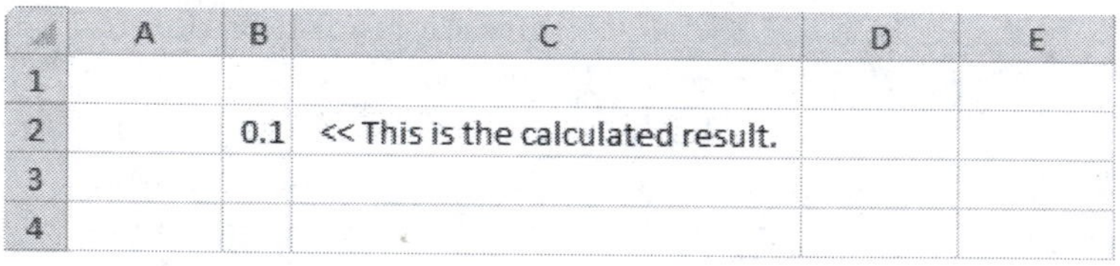

**Figure 2.60**
After adjusting the column width to fit the column contents.

| ▲ | A | B | C | D | E |
|---|---|---|---|---|---|
| 1 | | | | | |
| 2 | | 0.1 | << This is the calculated result. | | |
| 3 | | | | | |
| 4 | | | | | |

An alternative to Step 3 is to select **AutoFit Column Width** from the Cell Size menu shown in Figure 2.53. (**Home tab → Cells group → Format drop-down menu**.)

## 2.5 WORKING WITH WORKSHEETS

By default, an Excel *workbook* contains three worksheets, named Sheet1, Sheet2, and Sheet3. Sometimes you will need more than three worksheets, and usually the worksheets should be given more descriptive names.

There are some common modifications that apply to entire worksheets:

- Inserting new worksheets into a workbook.
- Renaming worksheets.
- Changing the color of the *worksheet tab*.
- Hiding a worksheet.
- Locking a worksheet.

### 2.5.1 Inserting New Worksheets into a Workbook

Inserting a new worksheet into a workbook is a very common task, and there are several ways to do it:

1. Use the keyboard shortcut [Shift + F11].
2. Click the **Insert Worksheet** button at the right end of the worksheet tabs (see Figure 2.61).
3. Use Ribbon options **Home tab → Cells group → Insert drop-down menu** and choose the **Insert Sheet** option.
4. Right-click on any worksheet tab and select **Insert . . . .**

**Figure 2.61**

The Insert Worksheet button next to the Sheet tabs.

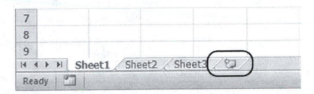

### 2.5.2 Renaming a Worksheet

A workbook can easily grow into a collection of dozens of worksheets. It is helpful to distinguish among worksheets by giving them meaningful names. There are three ways to *rename* a worksheet:

1. Double-click on the worksheet tab.
2. Right-click on the worksheet tab and select **Rename** from the pop-up menu.
3. Use Ribbon options **Home tab → Cells group → Format drop-down menu** and choose the **Rename Sheet** option.

### 2.5.3 Changing the Color of the Worksheet Tab

You can also make specific worksheets easier to identify by changing the color of the worksheet's tab. There are two ways to change the color of a worksheet's tab:

1. Right-click on the worksheet tab and select **Tab Color** from the pop-up menu.
2. Use Ribbon options **Home tab → Cells group → Format drop down menu** and choose the **Tab Color** option.

With either method, you can then select the new tab color from a color palette.

### 2.5.4 Hiding a Worksheet

It is also possible to *hide* a worksheet. When a worksheet is hidden, it is not visible but all of the cells are still updated and recalculated whenever a change is made in the visible parts of the workbook. The usual reasons for hiding worksheets are:

- to get less-used calculations out of the way
- to try to keep others from messing with your calculations

There are two ways to hide a worksheet:

1. Right-click on the worksheet tab and select **Hide** from the pop-up menu.
2. Use Ribbon options **Home tab → Cells group → Format drop-down menu**, choose the **Hide & Unhide** menu and select the **Hide Sheet** option.

Once you have a hidden worksheet, you may want to reveal or "unhide" it again. The same two approaches are used:

1. Right-click on *any* worksheet tab and select **Unhide** from the pop-up menu.
2. Use Ribbon options **Home tab → Cells group → Format drop-down menu**, choose the **Hide & Unhide** menu and select the **Unhide Sheet** option.

A list of hidden worksheets will be displayed. Select the worksheet that you want to unhide.

### 2.5.5 Locking a Worksheet

A hidden worksheet is still available for editing; you just have to "unhide" it. If you really want to keep people from messing with your calculations, you need to lock and password *protect* the worksheet. Actually the process is as follows:

1. Unlock all cells that you want the user to have access to after the worksheet is password protected.
2. Password protect the worksheet.

It may seem odd that you have to specify which cells should be unlocked, but by default, *all* cells in a worksheet are marked as "locked." The reason you can edit the cells in a worksheet is that locked cells are still available for editing until the worksheet is password protected. Before you password protect a worksheet, you need to think about how it will be used and which cells need to be available for editing. You must unlock all of the cells that you want the user to have access to after the worksheet is password protected.

As an example, let's return to the worksheet that calculates the area of a triangle (Figure 2.62).

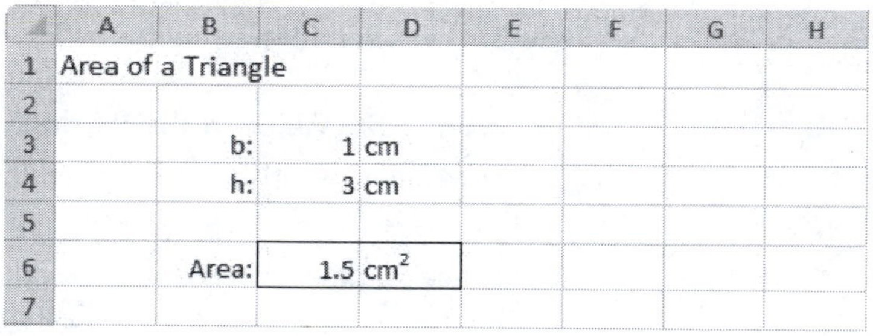

**Figure 2.62**
Example worksheet:
Calculating the area of a triangle.

The calculation is based on the values in cells C3 and C4, so those cells need to be unlocked before the worksheet is password protected. To unlock cells C3 and C4:

1. Select cells C3 and C4.
2. Use the Lock Cell toggle button from the Ribbon (**Home tab → Cells group → Format drop-down menu → Lock Cell toggle button**) to unlock the selected cells (Figure 2.63).

By unlocking cells C3 and C4, they will be available to a user after the worksheet is password protected.

There are two ways to password protect the worksheet:

1. Right-click on the worksheet tab and select **Protect Sheet . . .** from the pop-up menu.
2. Use Ribbon options **Home tab → Cells group → Format drop-down menu → Protect Sheet . . . .**

**Figure 2.63**
Using the Lock Cell toggle button to unlock selected cells.

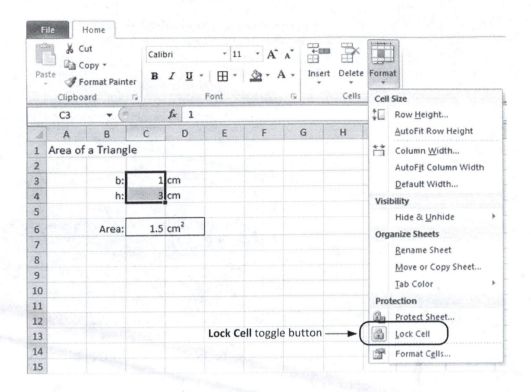

Either way, the Protect Sheet dialog (Figure 2.64) will open. Be sure that the **Protect worksheet and contents of locked cells** checkbox is checked, and enter a password. Remember the password because you will need to know it if you ever need to remove the password protection.

**Figure 2.64**
The Protect Sheet dialog box.

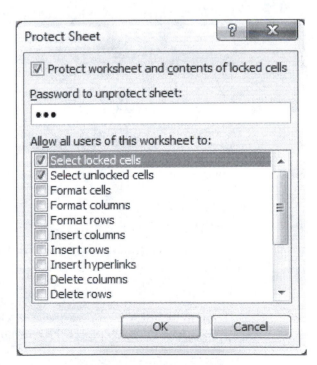

There are a variety of options that you can choose when protecting a worksheet. By default, users can still select both locked and unlocked cells, but they cannot change the contents of locked cells.

When you click the **OK** button on the Protect Sheet dialog, you will be asked to confirm the password (Figure 2.65).

**Figure 2.65**
Confirm Password
dialog box.

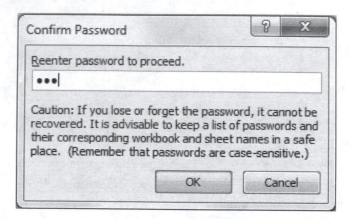

Once you have password protected the worksheet, only those cells that were unlocked are available to the user.

## 2.6 FORMATTING A DATA SET AS AN EXCEL TABLE

A table is simply a collection of information arranged into a rectangular grid. Any range of cells in a worksheet can be used to hold a table. For example, temperature data from several monitoring stations can be collected in tabular form, as shown in Figure 2.66.

**Note:** This is not weather data; it gets hot in Miami, but not 72.6°C (163°F).

**Figure 2.66**
Tabular data ready to be
made into an Excel table.

| | A | B | C | D |
|---|---|---|---|---|
| 1 | Station | Temp. (°C) | | |
| 2 | Newark | 12.5 | | |
| 3 | Santa Fe | 23.4 | | |
| 4 | Des Moines | 53.2 | | |
| 5 | Miami | 72.6 | | |
| 6 | Seattle | 23.5 | | |
| 7 | Phoenix | 34.6 | | |
| 8 | | | | |

### 2.6.1 Creating an Excel Table

Tables are even more powerful in Excel 2010. To turn data in cells into a table, you format the data as a table. This is done as follows:

1.  Select any cell in the data set (see Figure 2.67).

    **Note:** If you do not have contiguous data (i.e., if there are empty cells in the data set), then you must select the entire data set (including headings).

**Figure 2.67**
Turning a data set into an
Excel table.

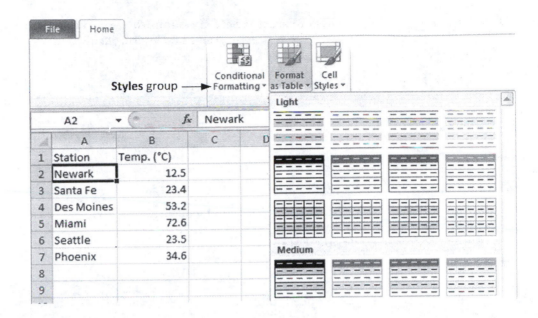

2. Use Ribbon options **Home tab → Styles group → Format as Table drop-down menu → Choose any style option**. (You can also create an Excel table from the Insert tab using these Ribbon options: **Insert tab → Tables group → Table button**. The table will be formatted using the Excel default table style.)

Excel will display the Format as Table dialog box (Excel 2007: Create Table dialog box), as shown in Figure 2.68. The dialog box provides an opportunity to confirm the cell range for the data set (A1:B7 in this example), and indicate whether or not there are headings included with your data. In this example, we do have headings (Station, Temp. (°C)).

**Figure 2.68**
The Create Table
dialog box.

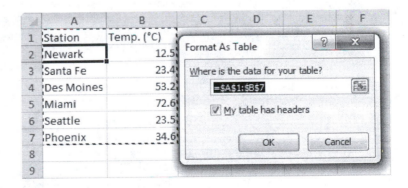

When you click the **OK** button on the Create Table dialog, Excel will convert the cell range to a table, and apply formatting. The result is shown in Figure 2.69.

Notice that the headings in cells A1 and B1 are now drop-down menus—this is the power of Excel tables. From the menus behind the headings you can *sort* or *filter* the data.

**Figure 2.69**
The Excel table.

| | A | B | C | D |
|---|---|---|---|---|
| 1 | Station | Temp. (°C) | | |
| 2 | Newark | 12.5 | | |
| 3 | Santa Fe | 23.4 | | |
| 4 | Des Moines | 53.2 | | |
| 5 | Miami | 72.6 | | |
| 6 | Seattle | 23.5 | | |
| 7 | Phoenix | 34.6 | | |
| 8 | | | | |

### 2.6.2 Sorting Data in Excel Tables

To arrange the data in alphabetical order by station, use the drop-down menu on the **Station** heading (Figure 2.70) and choose **Sort A to Z**. The result is shown in Figure 2.71.

**Figure 2.70**
Sorting the Excel table.

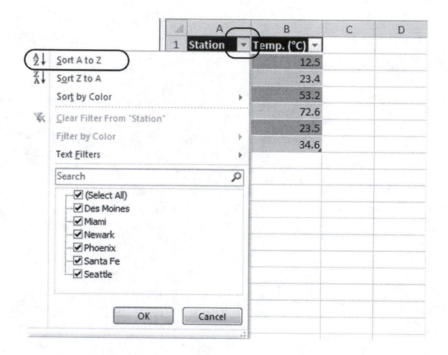

**Figure 2.71**
The table, sorted by Station.

| | A | B | C | D |
|---|---|---|---|---|
| 1 | Station | Temp. (°C) | | |
| 2 | Des Moines | 53.2 | | |
| 3 | Miami | 72.6 | | |
| 4 | Newark | 12.5 | | |
| 5 | Phoenix | 34.6 | | |
| 6 | Santa Fe | 23.4 | | |
| 7 | Seattle | 23.5 | | |
| 8 | | | | |

The drop-down menu indicator on the Station heading (see Figure 2.71) now has a small arrow pointing up; this indicates that the table has been sorted by station in ascending order.

Notice that Miami still has the highest temperature. When the stations were sorted, the rows were rearranged but the data on each row was kept together.

### 2.6.3 Filtering Data in Excel Tables

We can also filter the data in tables to see only the portion of the data set that meets the filter criteria. For example, we might filter the data to see only the stations that are reporting temperatures greater than 50°C. To apply the filter, use the drop-down menu on the **Temp. (°C)** heading. Then choose **Number Filters** and **Greater Than . . .** from the menus (see Figure 2.72). The Custom AutoFilter dialog box will open (Figure 2.73).

**Figure 2.72**
Filtering the data in an Excel table.

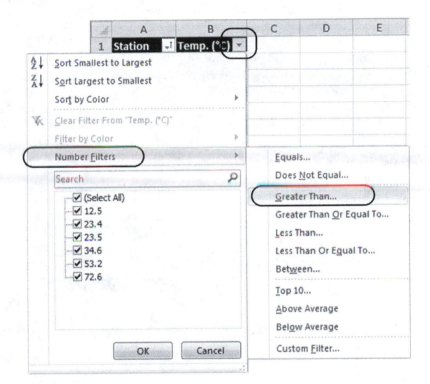

**Figure 2.73**
Custom AutoFilter dialog.

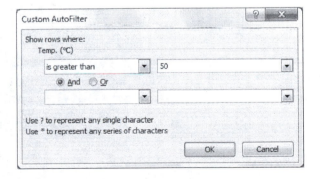

Enter the filter value (50 in this example) into the dialog box. Click **OK** to apply the filter. The result is shown in Figure 2.74. Notice that the drop-down menu indicator in cell B1 now includes a small funnel icon; this visually indicates that the table has been filtered on the temperature values. Also note (in Figure 2.74) that rows 5 through 7 are not visible. The rows containing temperature less than or equal to 50°C have been hidden (not deleted).

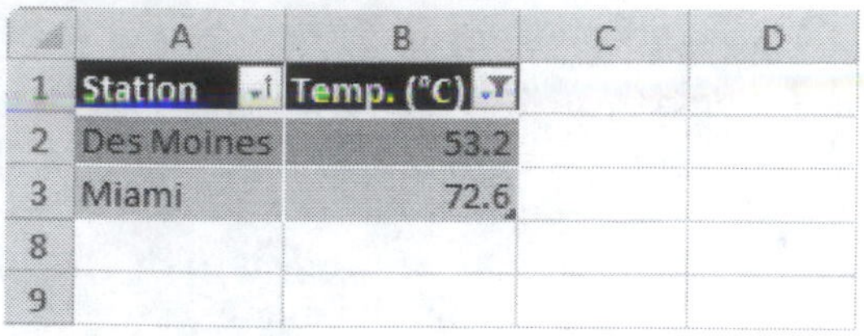

**Figure 2.74**
The filtered data.

To remove the filter, use the drop-down menu on the **Temp. (°C)** heading and choose **Clear Filter from Temp. (°C)**.

### 2.6.4 Using a Total Row with Excel Tables

Excel will automatically add a row to the bottom of a table that can be used to show the totals in each column—but a *Total Row* can be used for a lot more than just showing totals.

To add a Total Row, right-click anywhere on the table and select **Table → Totals Row** from the pop-up menu. The result is shown in Figure 2.75.

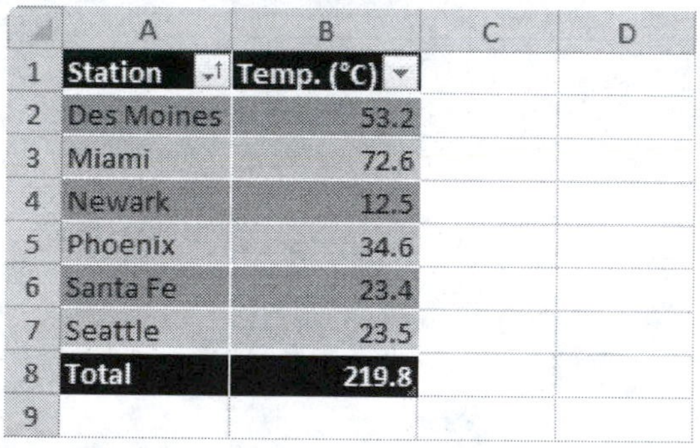

**Figure 2.75**
The example table with added Total Row.

There is little point in summing the temperatures from the various stations, but the Total Row can be used in other ways. If you click on cell B8 (the total), a drop-down menu handle will appear, as shown in Figure 2.76.

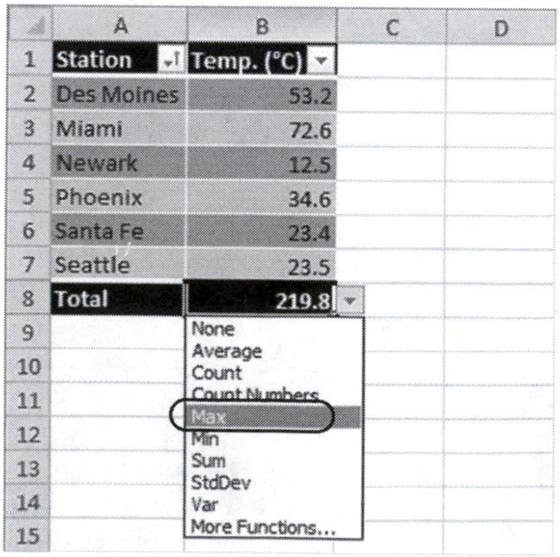

**Figure 2.76**
Options for the table's Total Row.

If you open the drop-down menu (as shown in Figure 2.76) you see that the Total Row can be used to calculate any of the following:

- Sum (or total)
- Average
- Count
- Maximum
- Minimum
- Standard Deviation
- Variance

If those don't meet your needs, you can use the **More Functions . . .** option.

## PRACTICE!

Suppose that you have a data set that includes melting and boiling points of various gases, as shown in Figure 2.77.

**Figure 2.77**

Gas property data.

| | A | B | C | D |
|---|---|---|---|---|
| 1 | Symbol | Property | Temp. (°C) | |
| 2 | He | Boil | -268.9 | |
| 3 | Xe | Melt | -112 | |
| 4 | Ne | Melt | -248.6 | |
| 5 | Ar | Boil | -185.8 | |
| 6 | Rn | Boil | -61.8 | |
| 7 | Ar | Melt | -189.3 | |
| 8 | Kr | Boil | -152.9 | |
| 9 | Kr | Melt | -157 | |
| 10 | Rn | Melt | -71 | |
| 11 | Ne | Boil | -245.9 | |
| 12 | | | | |

You want to sort the table so that:

- All of the melting points appear first and the boiling points appear last.
- The temperatures, within each category, should be listed in descending order.

This can be accomplished as follows:

1. Click anywhere inside the data set.
2. Create an Excel table from the data using Ribbon options: **Home tab →
   Styles group → Format as Table drop-down menu → Choose a style
   option**.
3. Verify the cell range containing the data and indicate that the table has headings in the Create Table dialog, as shown in Figure 2.78. The result is shown in Figure 2.79.
4. Use the drop-down menu on the **Temp (°C)** heading and select **Sort
   Largest to Smallest**.
5. Use the drop-down menu on the **Property** heading and select **Sort A to Z**.

The result is shown in Figure 2.80.

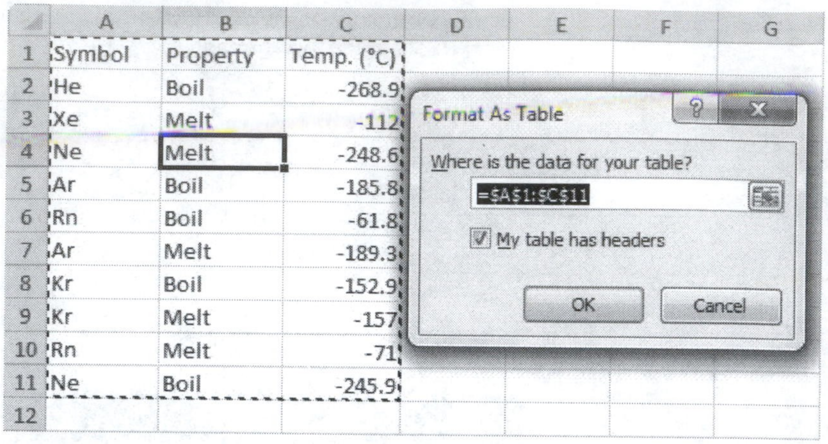

|   | A | B | C | D | E | F | G |
|---|---|---|---|---|---|---|---|
| 1 | Symbol | Property | Temp. (°C) | | | | |
| 2 | He | Boil | -268.9 | | | | |
| 3 | Xe | Melt | -112 | | | | |
| 4 | Ne | Melt | -248.6 | | | | |
| 5 | Ar | Boil | -185.8 | | | | |
| 6 | Rn | Boil | -61.8 | | | | |
| 7 | Ar | Melt | -189.3 | | | | |
| 8 | Kr | Boil | -152.9 | | | | |
| 9 | Kr | Melt | -157 | | | | |
| 10 | Rn | Melt | -71 | | | | |
| 11 | Ne | Boil | -245.9 | | | | |
| 12 | | | | | | | |

Format As Table

Where is the data for your table?

=$A$1:$C$11

☑ My table has headers

OK  Cancel

**Figure 2.78**
Creating an Excel table from a data set.

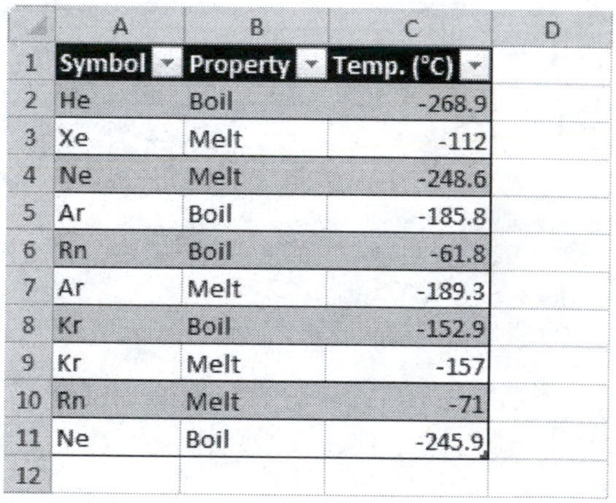

**Figure 2.79**
The Excel table.

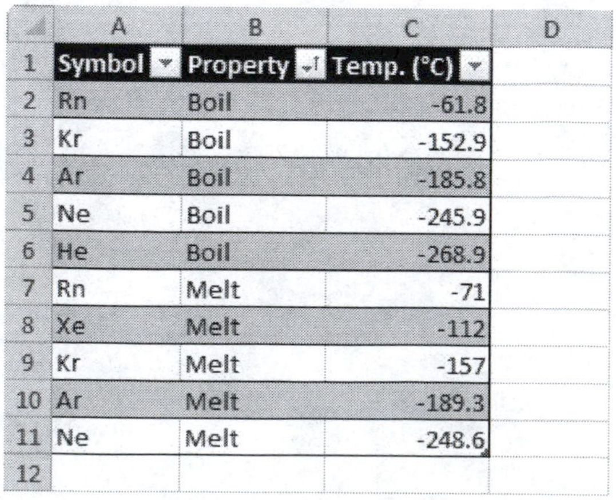

|   | A | B | C | D |
|---|---|---|---|---|
| 1 | Symbol ▾ | Property ▾ | Temp. (°C) ▾ | |
| 2 | He | Boil | -268.9 | |
| 3 | Xe | Melt | -112 | |
| 4 | Ne | Melt | -248.6 | |
| 5 | Ar | Boil | -185.8 | |
| 6 | Rn | Boil | -61.8 | |
| 7 | Ar | Melt | -189.3 | |
| 8 | Kr | Boil | -152.9 | |
| 9 | Kr | Melt | -157 | |
| 10 | Rn | Melt | -71 | |
| 11 | Ne | Boil | -245.9 | |
| 12 | | | | |

**Figure 2.80**
The sorted Excel table.

|   | A | B | C | D |
|---|---|---|---|---|
| 1 | Symbol ▾ | Property ▾↑ | Temp. (°C) ▾ | |
| 2 | Rn | Boil | -61.8 | |
| 3 | Kr | Boil | -152.9 | |
| 4 | Ar | Boil | -185.8 | |
| 5 | Ne | Boil | -245.9 | |
| 6 | He | Boil | -268.9 | |
| 7 | Rn | Melt | -71 | |
| 8 | Xe | Melt | -112 | |
| 9 | Kr | Melt | -157 | |
| 10 | Ar | Melt | -189.3 | |
| 11 | Ne | Melt | -248.6 | |
| 12 | | | | |

If, sometime later, you wanted to see only the boiling point values for argon and krypton, you could filter the table as follows:

6. Use the drop-down menu on the **Symbol** heading and check the boxes for the **Ar** and **Kr** Text Filters as shown in Figure 2.81.

**Figure 2.81**
Setting the Symbol filter values.

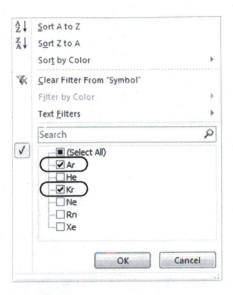

7. Use the drop-down menu on the **Property** heading and check the **Boil** box as shown in Figure 2.82.

**Figure 2.82**
Setting the Property filter value.

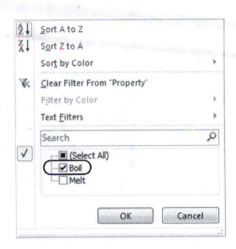

The filtered table is shown in Figure 2.83.

**Figure 2.83**
The result after filtering.

| | A | B | C | D |
|---|---|---|---|---|
| 1 | Symbol | Property | Temp. (°C) | |
| 3 | Kr | Boil | -152.9 | |
| 4 | Ar | Boil | -185.8 | |
| 12 | | | | |

## 2.7 CONDITIONAL FORMATTING

Formatting options may be applied conditionally. The term *conditional formatting* means you may choose logical criteria to format cells.

For example, consider again the temperature data shown in Figure 2.84. Suppose that we want to emphasize high and low temperatures. We can use conditional formatting to make the cells with the temperature extremes stand out.

**Figure 2.84**
Temperatures at various
monitoring locations.

| | A | B | C |
|---|---|---|---|
| 1 | Station | Temp. (°C) | |
| 2 | Newark | 12.5 | |
| 3 | Santa Fe | 23.4 | |
| 4 | Des Moines | 53.2 | |
| 5 | Miami | 72.6 | |
| 6 | Seattle | 23.5 | |
| 7 | Phoenix | 34.6 | |
| 8 | | | |

Create the worksheet shown in Figure 2.84. To initiate conditional formatting, select the cell range containing the temperature values, cells B2:B7, then click the Conditional Formatting button on the Ribbon [**Home tab → Styles group → Conditional Formatting button**]. A menu of options will be displayed as shown in Figure 2.85.

**Figure 2.85**
Applying conditional
formatting to cells B2:B7.

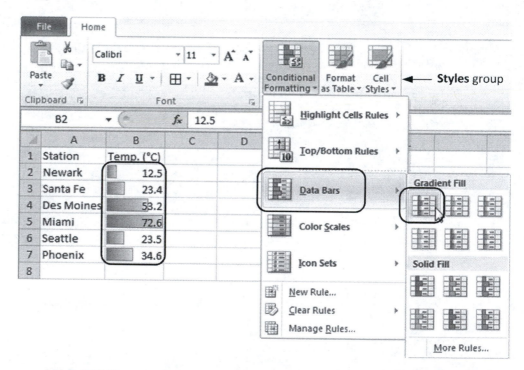

Excel 2007 and 2010 have several new options for conditional formatting, including the data bars used in Figure 2.85 and a feature new to Excel 2010 called a *sparkline* (a mini-chart in a cell). When you just want to visualize what's happening in your data, the new formats work well. But you can still apply your own rules to highlight data values meeting your own criteria.

For example, to highlight all cells containing temperatures greater than 50°C, you would create a rule as follows:

1. Select the cells containing the data to be examined; cells B2:B7 in this example.
2. Open the Greater Than dialog box using Ribbon options **Home tab → Styles group → Conditional Formatting drop-down menu → Highlight Cells Rules option → Greater Than option**. This string of menu options is illustrated in Figure 2.86. The dialog box is shown in Figure 2.87.

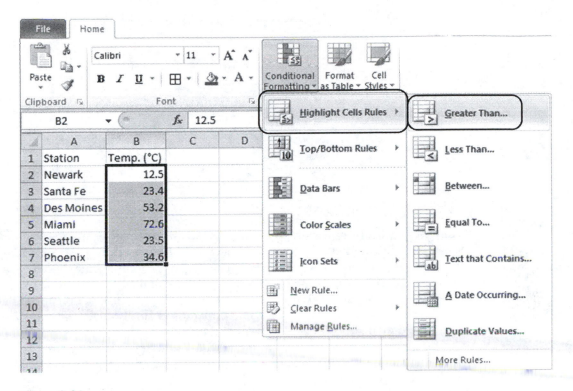

**Figure 2.86**
Ribbon options for creating a "greater than" rule.

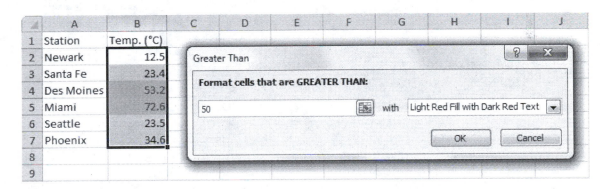

**Figure 2.87**
The Greater Than dialog box.

**3.** To highlight cells containing temperature values greater than 50°C, enter a 50 in the condition field, as shown in Figure 2.87. Then select a highlight format from the drop-down list on the right side of the dialog box, or select **Custom Format . . .** from the list to specify unique formatting for the highlighted cells.

The result is shown in Figure 2.88.

**Figure 2.88**
The result of applying a
conditional formatting rule
to highlight temperatures
above 50°C.

| | A | B | C |
|---|---|---|---|
| 1 | Station | Temp. (°C) | |
| 2 | Newark | 12.5 | |
| 3 | Santa Fe | 23.4 | |
| 4 | Des Moines | 53.2 | |
| 5 | Miami | 72.6 | |
| 6 | Seattle | 23.5 | |
| 7 | Phoenix | 34.6 | |
| 8 | | | |

APPLICATION

## PROCESS MONITORING

A lot of engineers, especially young engineers, find themselves in jobs where they are responsible for keeping an eye on how a manufacturing facility is functioning—for ensuring that equipment is operating correctly, routine maintenance is getting done, and quality control goals are being met. A lot of data is collected in manufacturing facilities to allow this type of monitoring to take place, so wading through the data to check for problems can be a chore. Because Excel is such a universally available program, many companies provide this monitoring data as an Excel workbook. So, the engineer comes to work in the morning and opens the workbook in Excel to see how well things went overnight. Figure 2.89 illustrates what a very small daily data set might look like.

| | A | B | C | D | E | F |
|---|---|---|---|---|---|---|
| 1 | Date: | Tuesday, October 20 | | | | |
| 2 | | Temperature | Speed | Power | Precision | Concentration |
| 3 | Units: | °C | RPM | Watt | mm | ppm |
| 4 | Min: | 45 | 1195 | 1.12 | 1.28 | 72 |
| 5 | Target: | 47 | 1200 | 1.23 | 1.30 | 75 |
| 6 | Max: | 52 | 1205 | 1.35 | 1.3 | 76 |
| 7 | Time | Temperature | Speed | Power | Precision | Concentration |
| 8 | 5:00 PM | 45 | 1200 | 1.11 | 1.22 | 75 |
| 9 | 5:05 PM | 45 | 1198 | 1.33 | 1.22 | 75 |
| 10 | 5:10 PM | 51 | 1197 | 1.10 | 1.30 | 75 |
| 11 | 5:15 PM | 44 | 1196 | 1.39 | 1.32 | 75 |
| 12 | 5:20 PM | 50 | 1200 | 1.31 | 1.36 | 75 |
| 13 | 5:25 PM | 47 | 1201 | 1.34 | 1.29 | 75 |
| 14 | 5:30 PM | 47 | 1200 | 1.05 | 1.33 | 75 |
| 15 | 5:35 PM | 44 | 1202 | 1.37 | 1.24 | 75 |
| 16 | 5:40 PM | 51 | 1198 | 1.38 | 1.34 | 75 |
| 17 | 5:45 PM | 50 | 1194 | 1.40 | 1.26 | 75 |
| 18 | 5:50 PM | 46 | 1201 | 1.14 | 1.35 | 75 |
| 19 | 5:55 PM | 45 | 1195 | 1.31 | 1.36 | 75 |
| 20 | 6:00 PM | 46 | 1196 | 1.08 | 1.28 | 75 |

**Figure 2.89**
Sample process monitoring data.

A quick look at the data shows a couple of temperatures that are below the 45°C minimum (5:15, 5:35), but finding problems would be easier if the values that are too low or too high were highlighted. Conditional formatting can make these values easy to spot.

The rule we will apply will instruct Excel to highlight values that are "not between" the minimum and maximum values listed in rows 4 and 6. You create a "not between" rule for the temperature values as follows:

1. Select the temperature measurements, cells B8:B20.
2. Open the New Formatting Rule dialog box using Ribbon options **Home tab → Styles group → Conditional Formatting drop-down menu → New Rule . . . option**. The dialog box is shown in Figure 2.90.

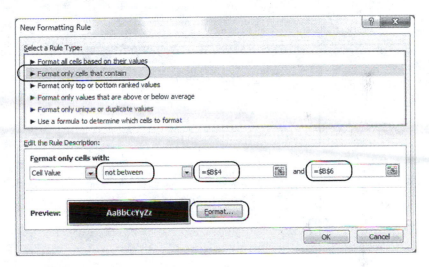

**Figure 2.90**
New Formatting Rule dialog box.

3. Choose the type of rule: "Format only cells that contain."
4. Since we want to highlight cells that are outside the desired temperature range, select "not between" as the test condition.
5. Indicate the minimum (45°C in cell B4) and maximum (52°C in cell B6) temperatures.
6. Click the **Format . . .** button to specify the format that should be used for the cells that are outside the desired temperature range. In this example they have been highlighted by using a dark background with bold, white text.

When you click the **OK** button, the rule will be applied to the selected cells and values outside the desired temperature range will be highlighted, as illustrated in Figure 2.91.

The same approach could be used to apply conditional formatting to the other types of measurements.

| | A | B | C | D | E | F |
|---|---|---|---|---|---|---|
| 1 | Date: Tuesday, October 20 | | | | | |
| 2 | | Temperature | Speed | Power | Precision | Concentration |
| 3 | Units: | °C | RPM | Watt | mm | ppm |
| 4 | Min: | 45 | 1195 | 1.12 | 1.28 | 72 |
| 5 | Target: | 47 | 1200 | 1.23 | 1.30 | 75 |
| 6 | Max: | 52 | 1205 | 1.35 | 1.3 | 76 |
| 7 | Time | Temperature | Speed | Power | Precision | Concentration |
| 8 | 5:00 PM | 45 | 1200 | 1.11 | 1.22 | 75 |
| 9 | 5:05 PM | 45 | 1198 | 1.33 | 1.22 | 75 |
| 10 | 5:10 PM | 51 | 1197 | 1.10 | 1.30 | 75 |
| 11 | 5:15 PM | **44** | 1196 | 1.39 | 1.32 | 75 |
| 12 | 5:20 PM | 50 | 1200 | 1.31 | 1.36 | 75 |
| 13 | 5:25 PM | 47 | 1201 | 1.34 | 1.29 | 75 |
| 14 | 5:30 PM | 47 | 1200 | 1.05 | 1.33 | 75 |
| 15 | 5:35 PM | **44** | 1202 | 1.37 | 1.24 | 75 |
| 16 | 5:40 PM | 51 | 1198 | 1.38 | 1.34 | 75 |
| 17 | 5:45 PM | 50 | 1194 | 1.40 | 1.26 | 75 |
| 18 | 5:50 PM | 46 | 1201 | 1.14 | 1.35 | 75 |
| 19 | 5:55 PM | 45 | 1195 | 1.31 | 1.36 | 75 |
| 20 | 6:00 PM | 46 | 1196 | 1.08 | 1.28 | 75 |

**Figure 2.91**
The process monitoring data after using conditional formatting to highlight temperatures outside desired operating conditions.

## KEY TERMS

Accounting format
AutoFit (column width)
Cell Border
Column Width
Conditional Format
Currency Format
Date format
Date-Time value
Fill Handle
Filter
Font Size
Font Style

General format
Hide Worksheet
Home Tab
Horizontal Alignment
Merge and Center
Protect (Lock)
  Worksheet
Rename Worksheet
Ribbon
Row Height
Scientific format
Series

Sort
Style
Table
Text format
Thousands separator
Time format
Vertical Alignment
Workbook
Worksheet
Worksheet Tab

# SUMMARY

**Common Data Formats**

- *General format*—Excel will choose a format on the basis of the contents of the cell. It is the default format for cells containing numbers or text.
- *Currency format*—Excel will display the numeric data with a currency symbol.
- *Accounting format*—Excel will display the numeric data with a currency symbol aligned at the left side of the cell.
- *Date format*—Excel will display the data as a date.
- *Scientific format*—Excel will display the number in scientific notation.
- *Text format*—The contents of the cell will be treated as text.
- *Time format*—Excel will display the data as a time.

*Assigning a Data Format to a Cell*

Use Ribbon options **Home tab** ➔ **Number group** ➔ **Number format drop-down list**.

**Date-Time Values**

The digits to the left of the decimal point are used to indicate the number of days since the starting date. The digits to the right of the decimal point are used to indicate the time as fraction of 24 hours.

Starting date defaults:

- PCs: January 1, 1900
- Macintosh computers: January 1, 1904

Changing the starting date on a PC: **File tab** (Office button in Excel 2007) ➔ **Options button** ➔ **Advanced group** ➔ **When Calculating this Workbook** section ➔ **Use 1904 Date System** check box.

**Fill Handle**

Use to quickly copy the contents of a cell across a column or row.

**Creating a Linear Series**

- Enter the first two series values in adjacent cells.
- Select the cells.
- Drag the fill handle with the left mouse button to create the series.

Drag the Fill Handle with the right mouse button for additional series options.

**Formatting Cells**

Most cell formatting options are located on the Ribbon's Home tab.

**Home Tab** ➔ **Font Group**

- Font style and size
- Bold, Italic, and Underline toggle buttons
- Cell Border (button and drop-down menu)
- Cell Fill Color (button and drop-down menu)
- Cell Font Color (button and drop-down menu)

**Home Tab** ➔ **Alignment Group**

- Vertical Alignment toggle buttons
- Horizontal Alignment toggle buttons
- Merge and Center button

**Home Tab → Number Group**

- Number format selector (drop-down list)
- Currency format toggle button and drop-down menu
- Percentage format toggle button
- Thousands Separator toggle button
- Increase and Decrease displayed digits buttons

**Adjusting Row Height** (three options)

- From a dialog box: **Home → Cells → Format drop-down menu → Row Height . . . button.**
- Drag the edge of the row heading with the mouse.
- Double-click the edge of the row heading to AutoFit the row height.

**Adjusting Column Width** (three options)

- From a dialog box: **Home → Cells → Format drop-down menu → Column Width . . . button.**
- Drag the edge of the column heading with the mouse.
- Double-click the edge of the column heading to AutoFit the column width.

**Worksheet Options**

- Insert a new worksheet: Right-click on any worksheet tab and select **Insert . . .**
- Rename a worksheet: Double-click on the worksheet tab.
- Change the color of the worksheet tab: Right-click on the worksheet tab and select **Tab Color**.
- Hide a worksheet: Right-click on the worksheet tab and select **Hide**.

**Locking a Worksheet**

1. Unlock any cells that need to be available after locking the worksheet.
   - Select cells to be unlocked.
   - Click the Lock Cell toggle button [**Home tab → Cells group → Format drop-down menu → Lock Cell toggle button**].
2. Lock the worksheet by password protecting the worksheet.
   - Right-click on the worksheet tab and select **Protect Sheet . . .**
   - Enter and confirm your chosen password.

**Excel Tables**

*Formatting a Data Set as an Excel Table*

1. Select any cell in the data set.
2. Use Ribbon options **Home tab → Styles group → Format as Table drop-down menu → Choose any style option** to create the table.

### Sorting Data

- Use the drop-down menu on the column heading.
- Select the sort option (ascending or descending) from the menu.

### Filtering Data

- Use the drop-down menu on the column heading.
- Choose the filter criteria from the menu.

### Using a Total Row

- Insert the total row: Right-click on the table and select **Table → Total Row**.
- Click on a cell in the total row to see the drop-down list icon.
- Use the drop-down list to select the function applied to the total row cell.

## Conditional Formatting

Used to highlight cells that meet certain criteria.

1. Select the cells that are to be assessed for conditional formatting.
2. Select or create a conditional format rule: **Home tab → Styles group → Conditional Formatting button.**

## PROBLEMS

1. When you use the General format, Excel attempts to interpret the type by your entry. Open a new workbook and type in the following items:

   1/2 _____

   1 1/2 _____

   $10.32 _____

   3E2 _____

   To what types does Excel automatically convert your text? To see the type, select the cell (with the data already entered) and then look on the Ribbon's **Home tab → Number group**. The drop-down list that is used to select a number format is also used to indicate the format applied to the currently selected cell.

2. Use the online help features of Excel to determine how Excel dealt with the Y2K problem. Assume that you are entering dates of birth in an Excel spreadsheet. If you enter 5/23/19, does Excel record the year to be 1919 or 2019? What about a date entered as 11/14/49?

3. Excel stores numbers with 15 digits of precision. Prove to yourself these limits of numerical precision.

   First, select an empty cell and format the cell to the Number format with 20 decimal places. (Choose the Number format from the Ribbon [**Home tab → Number group**], then right-click on the cell and choose **Format Cells . . .** from the pop-up menu to set the number of decimal places to 20.)

   Then, type the following formula into the cell. This formula instructs Excel to calculate the square root of 2:

   SQRT(2)

Since the square root of 2 is an irrational number, the fractional part of its decimal representation has an unending number of nonrepeating digits. At what number of digits does Excel's accuracy stop?

4. The table in Figure 2.92 shows the temperatures recorded at several monitoring stations for the months of January and February 2008. The data are in no particular order. Create a worksheet that looks like Figure 2.92. Convert the data into an Excel table, then sort the data in your worksheet by station and then by date in ascending order. The results should look like Figure 2.93. (The formatting of the table may be different, but the arrangement of the data values should be the same.)

**Figure 2.92**

Temperature data.

| | A | B | C | D |
|---|---|---|---|---|
| 1 | Station | Date | Temp (°C) | |
| 2 | A02 | 2/1/2008 | 23.5 | |
| 3 | A07 | 1/15/2008 | 14.6 | |
| 4 | A02 | 2/15/2008 | 0.5 | |
| 5 | B05 | 2/3/2008 | 20.0 | |
| 6 | B05 | 1/12/2008 | 34.3 | |
| 7 | C12 | 1/5/2008 | 20.2 | |
| 8 | A02 | 2/23/2008 | 19.6 | |
| 9 | B05 | 2/1/2008 | 22.3 | |
| 10 | | | | |

**Figure 2.93**

The sorted Excel table.

| | A | B | C | D |
|---|---|---|---|---|
| 1 | Station | Date | Temp (°C) | |
| 2 | A02 | 2/1/2008 | 23.5 | |
| 3 | A02 | 2/15/2008 | 0.5 | |
| 4 | A02 | 2/23/2008 | 19.6 | |
| 5 | A07 | 1/15/2008 | 14.6 | |
| 6 | B05 | 1/12/2008 | 34.3 | |
| 7 | B05 | 2/1/2008 | 22.3 | |
| 8 | B05 | 2/3/2008 | 20.0 | |
| 9 | C12 | 1/5/2008 | 20.2 | |
| 10 | | | | |

5. Create a worksheet that looks like Figure 2.92. Use conditional formatting to highlight temperatures less than 20.0°C. Your result should be very similar to Figure 2.94.

**Figure 2.94**

Unsorted data highlighting temperatures less than 20°C.

| | A | B | C | D |
|---|---|---|---|---|
| 1 | Station | Date | Temp (°C) | |
| 2 | A02 | 2/1/2008 | 23.5 | |
| 3 | A07 | 1/15/2008 | 14.6 | |
| 4 | A02 | 2/15/2008 | 0.5 | |
| 5 | B05 | 2/3/2008 | 20.0 | |
| 6 | B05 | 1/12/2008 | 34.3 | |
| 7 | C12 | 1/5/2008 | 20.2 | |
| 8 | A02 | 2/23/2008 | 19.6 | |
| 9 | B05 | 2/1/2008 | 22.3 | |
| 10 | | | | |

6. Create a worksheet containing the table of noble gases shown in Figure 2.77. Create an Excel table, and then sort the table so that the elements are listed in alphabetical order. For each element, place the melting point first, followed by the boiling point. Your results should look like Figure 2.95.

**Figure 2.95**

Sorted properties of the noble gases.

| | A | B | C | D |
|---|---|---|---|---|
| 1 | Symbol | Property | Temp. (°C) | |
| 2 | Ar | Melt | -189.3 | |
| 3 | Ar | Boil | -185.8 | |
| 4 | He | Boil | -268.9 | |
| 5 | Kr | Melt | -157.0 | |
| 6 | Kr | Boil | -152.9 | |
| 7 | Ne | Melt | -248.6 | |
| 8 | Ne | Boil | -245.9 | |
| 9 | Rn | Melt | -71.0 | |
| 10 | Rn | Boil | -61.8 | |
| 11 | Xe | Melt | -112.0 | |
| 12 | | | | |

7. Use conditional formatting to highlight the boiling points in the table of Figure 2.95 by applying conditional formatting to all cells containing the word "Boil." Your results should highlight the same items as in Figure 2.96.

**Figure 2.96**

Noble gas property table with boiling points highlighted.

| | A | B | C | D |
|---|---|---|---|---|
| 1 | Symbol | Property | Temp. (°C) | |
| 2 | Ar | Melt | -189.3 | |
| 3 | Ar | Boil | -185.8 | |
| 4 | He | Boil | -268.9 | |
| 5 | Kr | Melt | -157.0 | |
| 6 | Kr | Boil | -152.9 | |
| 7 | Ne | Melt | -248.6 | |
| 8 | Ne | Boil | -245.9 | |
| 9 | Rn | Melt | -71.0 | |
| 10 | Rn | Boil | -61.8 | |
| 11 | Xe | Melt | -112.0 | |
| 12 | | | | |

8. Create and format a table that looks like Figure 2.97.

**Figure 2.97**

Student grades.

| | A | B | C | D | E | F | G |
|---|---|---|---|---|---|---|---|
| 1 | Name | Quiz 1 | Quiz 2 | Midterm | Quiz 3 | Final | |
| 2 | Bob | 23 | 12 | 43 | 21 | 54 | |
| 3 | Maria | 32 | 10 | 40 | 26 | 55 | |
| 4 | Ralph | 14 | 12 | 34 | 20 | 45 | |
| 5 | Deepak | 24 | 13 | 38 | 22 | 58 | |
| 6 | | | | | | | |

9. The median is the middle value in a sorted data list. Create the data table shown in Figure 2.98. Create an Excel table, and then sort the data to determine the median grade for this group of students.

**Figure 2.98**
Grades

| | A | B |
|---|---|---|
| 1 | | |
| 2 | Student | Score |
| 3 | Sal | 87 |
| 4 | Sali | 94 |
| 5 | Sally | 72 |
| 6 | Sam | 82 |
| 7 | Sara | 88 |
| 8 | Sarah | 78 |
| 9 | Su Nee | 95 |
| 10 | Stewart | 91 |
| 11 | Stuart | 68 |
| 12 | Suri | 84 |
| 13 | | |

10. Ben received a notice from his movie rental store on May 7 informing him that he failed to return a movie. After finding the movie under the couch cushion, he noticed that the due date was March 17. Use Excel's ability to subtract dates to determine if it will be cheaper to pay the $1.29 per day late fee, or buy the movie for $69. If you set up your worksheet as shown in Figure 2.99, you can compute the number of late days by entering the formula

$$= C2 - C3$$

in cell C5.

**Figure 2.99**
Calculating the number of days between two dates.

| | A | B | C | D |
|---|---|---|---|---|
| 1 | | | | |
| 2 | | Notice Date: | 7-May | |
| 3 | | Due Date: | 17-Mar | |
| 4 | | | | |
| 5 | | Days Late: | | |
| 6 | | | | |

11. You can use Excel's Date format to find out on what day of the week an event occurred. To do so, enter the date in a cell on a worksheet, then choose and select a Type for the Date format that includes the day of the week. The date you entered will then be displayed along with the day of the week.

For example, say your friend Anna was born on December 28, 1984. You could find out she was born on a Friday by entering 12/28/1984 into a cell and then reformatting the cell to include the day of the week. The result is shown in Figure 2.100.

**Figure 2.100**
Long Date formatting used to determine the day of the week.

| | A | B | C |
|---|---|---|---|
| 1 | | | |
| 2 | | Friday, December 28, 1984 | |
| 3 | | | |
| 4 | | | |

Find out the day of the week for each of these historical events:

**a.** Stock market crash, October 29, 1929 (hint, it is known as Black Tuesday)

**b.** First moon landing, July 20, 1969

**c.** Cinco de Mayo (May 5), 2016

**d.** Your birthdate

**12.**   Use Excel to create a schedule for your classes:

**Step 1:**   Enter Monday and Tuesday in adjacent cells.

**Step 2:**   Select the cells containing Monday and Tuesday, and then use the Fill Handle to complete the headings for the class schedule. The result of this step is shown in Figure 2.101.

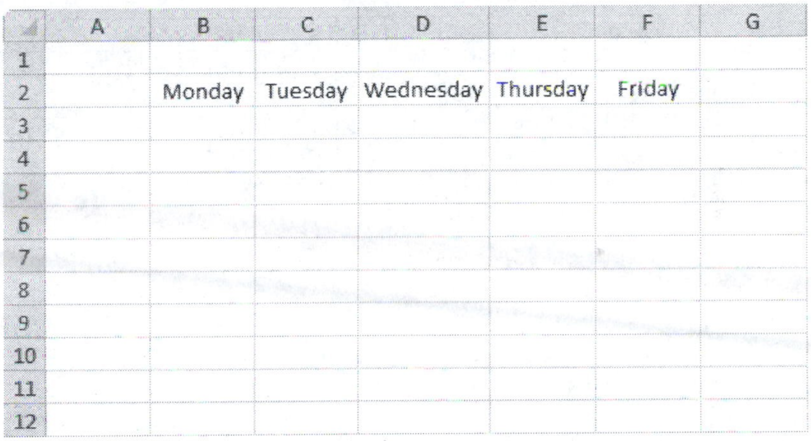

**Figure 2.101**
Day headings for a class schedule.

| | A | B | C | D | E | F | G |
|---|---|---|---|---|---|---|---|
| 1 | | | | | | | |
| 2 | | Monday | Tuesday | Wednesday | Thursday | Friday | |
| 3 | | | | | | | |
| 4 | | | | | | | |
| 5 | | | | | | | |
| 6 | | | | | | | |
| 7 | | | | | | | |
| 8 | | | | | | | |
| 9 | | | | | | | |
| 10 | | | | | | | |
| 11 | | | | | | | |
| 12 | | | | | | | |

**Step 3:**   Just below and to the left of the cell containing Monday, enter the time that your school's first class starts. Figure 2.102 assumes that classes begin at 8:00 a.m.

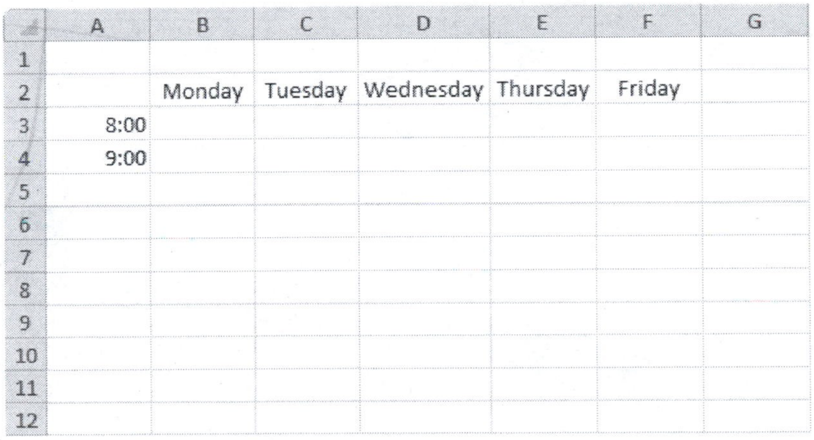

**Figure 2.102**
First two class starting times entered in column A.

| | A | B | C | D | E | F | G |
|---|---|---|---|---|---|---|---|
| 1 | | | | | | | |
| 2 | | Monday | Tuesday | Wednesday | Thursday | Friday | |
| 3 | 8:00 | | | | | | |
| 4 | 9:00 | | | | | | |
| 5 | | | | | | | |
| 6 | | | | | | | |
| 7 | | | | | | | |
| 8 | | | | | | | |
| 9 | | | | | | | |
| 10 | | | | | | | |
| 11 | | | | | | | |
| 12 | | | | | | | |

**Step 4:**   Just below the time entered in Step 3, enter the time that the second class starts.

**Step 5:**   Complete the remaining class start times in the time column. You may be able to use the Fill Handle to speed the process if your school has evenly spaced class times.

**Step 6:** Enter an abbreviation for each of your classes in the appropriate cell.

**Step 7:** Format the class schedule to make it easy to read. Figure 2.103 has been completed as an example.

**Figure 2.103**
Completed class schedule.

| | A | B | C | D | E | F | G |
|---|---|---|---|---|---|---|---|
| 1 | | | | | | | |
| 2 | | Monday | Tuesday | Wednesday | Thursday | Friday | |
| 3 | 8:00 | MATH | MATH | | MATH | MATH | |
| 4 | 9:00 | CHEM | | CHEM | | CHEM | |
| 5 | 10:00 | | | | | | |
| 6 | 11:00 | | HIST | | HIST | | |
| 7 | 12:00 | | HIST | | HIST | | |
| 8 | 1:00 | | | Chem Lab | | | |
| 9 | 2:00 | | | Chem Lab | | | |
| 10 | 3:00 | | | Chem Lab | | | |
| 11 | 4:00 | | | Chem Lab | | | |
| 12 | | | | | | | |

# 3

# Formulas and Functions

## Sections

## Objectives

*After reading this chapter, you should be able to perform the following tasks:*

- Refer to cells and cell ranges in a worksheet.
- Create formulas in a worksheet.
- Locate and use Excel's predefined functions.
- Use absolute and relative cell references in formulas and functions.
- Perform simple matrix operations with Excel.
- Debug worksheet formulas that contain errors.
- Record and run a macro.

## 3.1 INTRODUCTION

The ability to manipulate *formulas*, arrays, and mathematical functions is the most important feature of Excel for engineers. A common scenario is for an engineer to test and refine potential solutions to a problem by using Excel. When the engineer is satisfied that the solution works for small data sets, the solution might be translated to a programming language such as C or FORTRAN. The resulting program could then be executed on a powerful workstation or supercomputer, using large data sets. This use of a worksheet is called building a *prototype*. An application package such as Excel is useful for building prototypes because it allows solutions to be quickly developed and easily modified.

In these sections, the basics of using Excel to solve mathematical problems will be presented. This includes writing formulas in a cell based on values in other cells, as well as using Excel's predefined functions to solve problems. Some examples of typical engineering calculations that can be performed using Excel's predefined functions will also be presented, including statistical calculations, trigonometric calculations, and matrix calculations.

In addition, you will be introduced to Excel macros—a method for recording and executing a series of actions. The use of macros can be a time-saving feature as you learn to solve problems that require a series of computations.

## 3.2 REFERENCING CELLS AND CELL RANGES

Calculations in Excel typically use values stored in various cells to calculate new values. The formulas (or equations) used to calculate the new values must refer to, or reference, the values in the other cells. The reference can be to the value in a single cell or to the set of values in a range of cells. References frequently use the cell location (e.g., B1 or A7) to identify a specific cell, but cells or cell ranges can also be given a name, and this can make formulas easier to understand. This will be presented in more detail in the following sections.

### 3.2.1 Cell References

A cell can be referenced by using its column letter and row number. The selected cell in Figure 3.1 can be referred to as cell B3. Notice in Figure 3.1 that the *Name Box* at the left end of the *Formula bar* displays the cell reference for the currently selected cell.

**Note:** The Formula bar is displayed by default, but it can be turned off. In Excel 2007 or 2010, if the Formula bar is not displayed on your screen, use the following *Ribbon* options to cause it to be displayed: **View tab → Show/Hide group → check the Formula Bar box**. (In Excel 2003, choose View → Formula Bar from the main menu.)

**Note:** If the Formula bar is not being displayed when you open a new workbook, you can change the Excel 2007 default with these options: **File tab (Office button in Excel 2007) → Options button → Advanced tab → Display section →** check the box labeled **Show Formula Bar**.

**Figure 3.1**
Reference a cell by using its row and column designations.

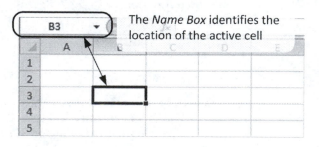

The *Name Box* identifies the location of the active cell

If a value (number or text) is entered into cell B3, we can reference the contents of cell B3. For example, if we enter a value of 12 into cell B3 (as shown in Figure 3.2), we can say "cell B3 contains the value 12." In this way we are referring to cell B3 to access the value, 12.

**Figure 3.2**

The Formula bar shows the contents of the currently selected cell.

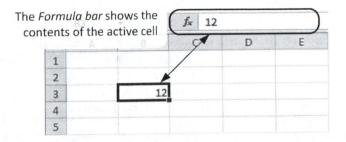

The *Formula bar* shows the contents of the active cell

If we put a simple formula into another cell, the formula can use the value of 12 by referencing B3. For example, if the formula = B3/2 is entered into cell B4, we can divide 12 by 2. The result is shown in Figure 3.3. Notice that the Formula bar displays the *contents* of cell B4 (the formula), while the *result* of the calculation is displayed in cell B4.

**Figure 3.3**

The formula in cell B4 references the value in cell B3.

|  | A | B | C | D | E |
|---|---|---|---|---|---|
| B4 | | | | fx | =B3/2 |
| 1 | | | | | |
| 2 | | | | | |
| 3 | | 12 | | | |
| 4 | | 6 | | | |
| 5 | | | | | |

### 3.2.2 Referencing a Range of Cells

Some calculations are performed on multiple values, for example, calculating a sum or an average. In Excel, multiple values are typically stored in adjacent cells. When the group of values is referenced, it is termed a **cell range**. A rectangular cell range is identified by the top-left and bottom-right cell references, separated by a colon (:). The cell range selected in Figure 3.4 is referenced as A2:B6.

**Figure 3.4**

Selected cell range A2:B6.

|  | A | B | C | D | E |
|---|---|---|---|---|---|
| 1 | | | | | |
| 2 | 1.4 | 1.5 | Terms: | | |
| 3 | 1.6 | 1.3 | • A2 is the *Active Cell* | | |
| 4 | 1.3 | 1.5 | • The selected *cell* | | |
| 5 | 1.5 | 1.7 | *range* is A2:B6 | | |
| 6 | 1.7 | 1.9 | | | |
| 7 | | | | | |

A simple Excel function that uses a range of values is the **SUM function**. If =SUM(A2:B6) is entered into cell D2, the sum of the values in the cell range A2:B6 will be computed and displayed in cell D2, as shown in Figure 3.5.

**Figure 3.5**

Example of a function (SUM) that references a cell range.

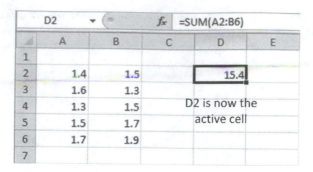

### 3.2.3 Naming a Cell or Range of Cells

It is often convenient to assign a name to a cell or range of cells. Once the name has been assigned, it can be used in formulas and functions to reference the cell or range of cells.

To assign a name to a range of cells, follow these steps:

1. Select the range of cells to be named (as shown in Figure 3.4).
2. Enter a name for the cell range in the Name box. In Figure 3.6, the range has been named MyValues.

**Figure 3.6**

The cell range A2:B6 has been named MyValues.

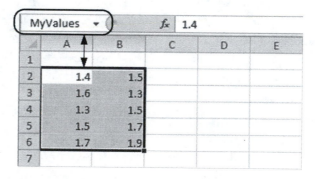

Once the range has been named, the name can be used in the *SUM* function in cell D2, as shown in Figure 3.7.

**Figure 3.7**

Example of using a named cell range in a formula.

Using names in formulas instead of row and column designations can make your worksheets easier to understand.

Once you have defined a *named cell* range, how do you "undefine" it if you need to? In Excel 2007 and 2010 you can use the **Name Manager**, which is available from the Ribbon's Formulas tab, as indicated in Figure 3.8.

**Figure 3.8**

Use the Name Manager to edit or delete a name.

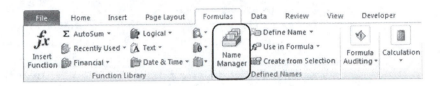

## 3.3 CREATING AND USING FORMULAS

A formula in Excel consists of a mathematical expression. For the most part, the expression is defined by using common mathematical operators:

+    Addition [+]

−    Subtraction [−]

\*    Multiplication [*], or [Shift + 8]

/    Division [/]

^    Exponentiation [^], or [Shift + 6]

A cell containing a formula can display either the formula definition or the formula's calculated result. The default is to display the results in the cell. This is usually preferable since the formula definition for the currently active cell is displayed in the Formula bar.

To display formulas in cells, use the following options: **Formulas tab → Formula Auditing group → Show Formulas** button. (In Excel 2003: Tools → Options → choose the View panel on the Options dialog box → check the box labeled Formulas in the Window options area.)

In Figure 3.9, a formula has been entered in cell D1. The formula bar shows the formula definition to be

$$= A1*B1/C1$$

**Figure 3.9**

An example of a formula in cell D1.

| | A | B | C | D | E |
|---|---|---|---|---|---|
| D1 | | | $f_x$ | =A1*B1/C1 | |
| 1 | 2.3 | 4.1 | 5.4 | 1.746296 | |
| 2 | | | | | |

Like all formulas, the formula in cell D1 starts with an equal sign (=). The equal sign tells Excel that the cell contains a formula and not text. The result of applying this formula is displayed in cell D1 as 1.746296.

### 3.3.1 Formula Syntax

An Excel formula uses a strict *syntax*, or set of rules that govern how formulas must be entered, so that Excel can correctly interpret them. This means that you must learn the rules for entering a formula and adhere strictly to those rules. There is some good news here: The rules are not very hard to learn.

**Rules:**

- A formula can consist of operators, predefined function names, cell references, and cell names.
- A formula always begins with an equal sign (actually, Excel will also allow you to use a plus symbol).

  The equal sign is an indicator for Excel to evaluate the expression that follows as a formula, instead of simply interpreting the contents in the cell as a text phrase. Try removing the equal sign from a formula and see what happens.
- You can enter a formula into the Formula window on the Formula bar, or directly into the destination cell.
- You have to be careful about *operator precedence*—this is discussed in the next section.

### 3.3.2 Arithmetic Operators and Operator Precedence

Table 3.1 lists Excel's arithmetic operators. The operators are listed in order of *precedence*. Precedence indicates which operators are evaluated first; an operator that has higher precedence will always be evaluated before an operator of lower precedence. When operators have the same level of precedence, such as addition and subtraction, the operators are evaluated from left to right. If a different order of evaluation is needed, you must use parentheses to ensure that your formulas evaluate correctly.

**Table 3.1 Precedence of Arithmetic Operators**

| Precedence | Operator | Operation |
|:---:|:---:|:---|
| 1 | % | Percentage |
| 2 | ^ | Exponentiation |
| 3 | *, / | Multiplication, Division |
| 4 | +, − | Addition, Subtraction |

Let's apply the rules of precedence to a few formulas, and calculate the expected results.

EXAMPLE 1: $= 8/2 + 2$

The operator of highest precedence is the division symbol, so the first operation is $8/2 = 4$. That result is used in the next operation, $4 + 2 = 6$. The result of this formula should be 6, and that is what Excel calculated in cell B1, shown in Figure 3.10.

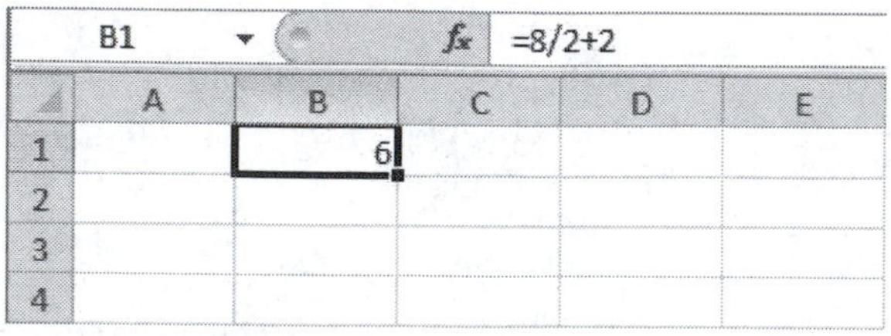

**Figure 3.10**

Excel's evaluation of the formula $= 8/2 + 2$.

EXAMPLE 2: $= 8/(2 + 2)$

In Example 2, the operation in parentheses must be evaluated first, so $2 + 2 = 4$. This result is used in the division, $8/4 = 2$. The result of this formula should be 2, and that is what Excel calculated in cell B2, shown in Figure 3.11.

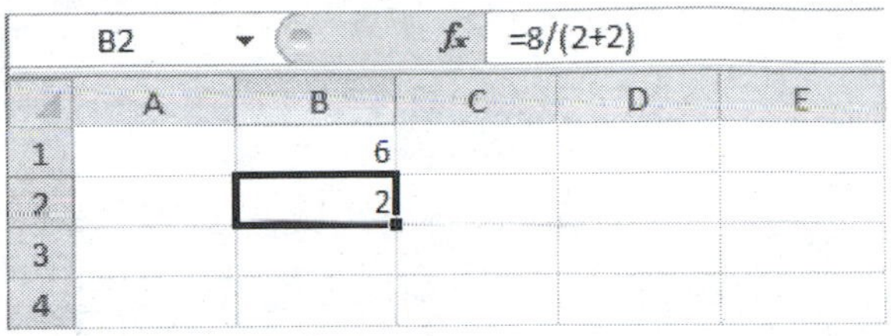

**Figure 3.11**

Excel's evaluation of the formula $= 8/(2+2)$.

EXAMPLE 3: $= 3\wedge 2 + 24/2/4 + 3$

The exponentiation has the highest precedence and is evaluated first:

$$= [9] + 24/2/4 + 3$$

The divisions have the next highest priority and will be evaluated from left to right:

$$= [9] + [12]/4 + 3$$
$$= [9] + [3] + 3$$

The additions are then evaluated from left to right:

$$= [12] + 3$$
$$= [15]$$

The result of this formula should be 15, and that is what Excel calculated in cell B3, shown in Figure 3.12.

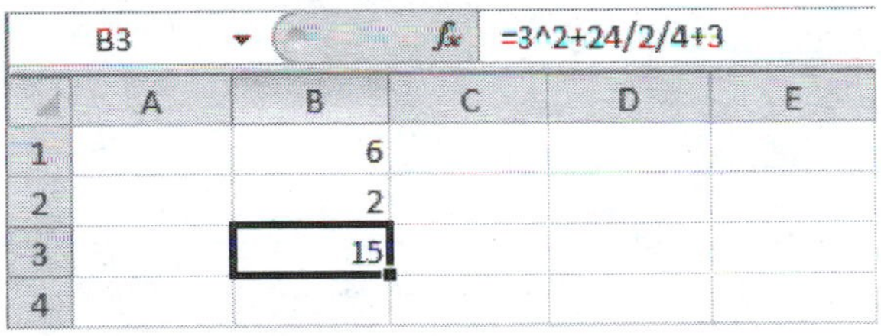

**Figure 3.12**

Excel's evaluation of the formula $= 3\wedge 2 + 24/2/4 + 3$.

There are other operators in Excel for operations like manipulating text and performing logical (Boolean) comparisons. Check out the online help system for precedence information on additional operators.

## PRACTICE!

Practice creating arithmetic expressions by following these instructions:

1. Select a cell.
2. Type the following arithmetic expression into the cell:

$$= 6/2 + 3$$

Try to figure out what the result will be before pressing the **Enter** key.

3. Do the same for the following expressions:

$$= 6/(2+3)$$
$$= 2\wedge 2 - 1$$
$$= 2\wedge(2-1)$$
$$= 2 + 4/2/2 - 4$$

*Answers:*

$$= 6/2 + 3 \qquad [6]$$
$$= 6/(2 + 3) \qquad [1.2]$$
$$= 2\char`^2 - 1 \qquad [3]$$
$$= 2\char`^(2 - 1) \qquad [2]$$
$$= 2 + 4/2/2 - 4 \qquad [-1]$$

## 3.4 USING EXCEL'S BUILT-IN FUNCTIONS

Excel has a large number of predefined or ***built-in functions*** available for immediate use. These functions are similar to functions in a programming language because they each require a specified number of arguments as input, and they each return a value. The built-in functions may be selected directly by choosing the ***Insert Function button*** from the Formula bar (indicated in Figure 3.13). When you click on the **Insert Function** button, the Insert Function dialog box will appear, as shown in Figure 3.14.

**Figure 3.13**
The Insert Function button is located on the Formula bar.

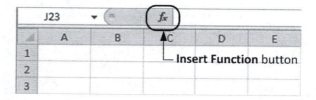

**Figure 3.14**
The Insert Function dialog box.

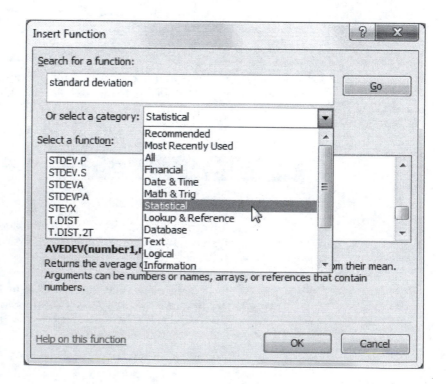

From the Insert Function dialog box, you can select functions in a variety of ways.

- *Search for a function:*   Type a topic in the box labeled **Search for a function**. The term "standard deviation" was used in the example shown in Figure 3.14.
- *Use categories:*   Select a category of functions from the drop-down menu, and then select a function from the **Select a function** list.
- *Reuse recently used functions:*   Once you have used several functions in a worksheet, you can quickly access them again by selecting the **Most Recently Used** item in the category drop-down list.

Most engineers will find lots of uses for functions from the **Math & Trig** and **Statistical** categories. Some engineers find functions in other categories very helpful as well. Several examples of available built-in functions will be presented later.

Once you have selected a function, the Function Arguments dialog box for that function will be displayed. As an example, the Function Arguments dialog box for the *SUM* function is shown in Figure 3.15.

**Figure 3.15**

Function Arguments dialog box for the *SUM* function.

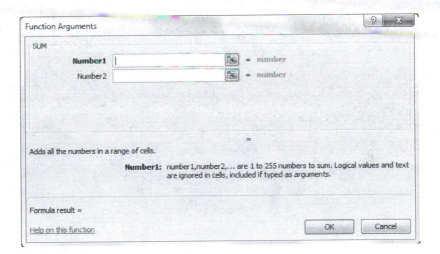

The *SUM* function calculates the sum of a series of values. The values must be passed into the function as a *function argument* or input. The *SUM* function typically takes a cell range as the only argument, but you can have multiple arguments, separated by commas, and the *SUM* function will add all of the input values. The **Number1** and **Number2** fields on the Function Arguments dialog box can hold a number, a reference to a single cell, or (most common) a reference to a range of cells. If you use the **Number1** and **Number2** fields on the dialog box, Excel will display additional fields so that you can sum additional values.

The most common use of the *SUM* function is to calculate the sum of a series of values stored in a cell range. The cell references must be passed into the *SUM* function.

Cell references can be entered as function arguments in one of two ways:

- A cell location or cell range can be typed into the formula.
- The cell or cell range can be selected by using the mouse.

Let's walk through an example, using the *SUM* function.

1. Type the values 7.5 and 6.2 into cells A3 and B3, respectively, as shown in Figure 3.16.

**Figure 3.16**
Example: Values to be summed.

| | A | B | C | D | E |
|---|---|---|---|---|---|
| 1 | | | | | |
| 2 | | | | | |
| 3 | 7.5 | 6.2 | | | |
| 4 | | | | | |

2. Select cell C3.
3. Choose the Insert Function button from the Formula bar. The Insert Function dialog box (Figure 3.14) will appear.
4. Choose the **SUM** function from the **Math & Trig** category.
5. The Function Arguments dialog box for the *SUM* function will appear, as shown in Figure 3.17.

**Figure 3.17**
The Function Arguments dialog for the *SUM* function with the Select from Worksheet button indicated.

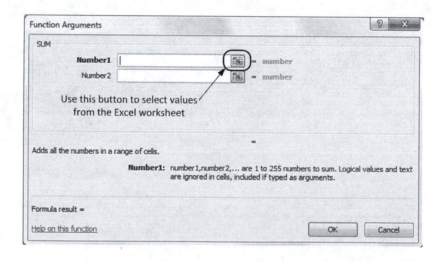

6. You have two choices at this point. You can type the input cell range (A3:B3) into the box labeled **Number1**, or you can click the small button at the right of the data entry field. This button is called the *Select from Worksheet* button, and it has been indicated in Figure 3.17. The button represents a worksheet grid with an arrow pointing to a selection, and the button is used to jump back to the worksheet so that you can select a range of cells with the mouse. For now,

click the Select from worksheet button. The Function Arguments dialog box will shrink so that the worksheet is visible, as shown in Figure 3.18. You can now select the input cell range (the values to be summed) by using the mouse. The selected range is indicated in the Function Arguments dialog box, in the cell containing the formula, and on the Formula bar (Figure 3.18).

**Figure 3.18**
Selecting the input cell range for the *SUM* function with a mouse.

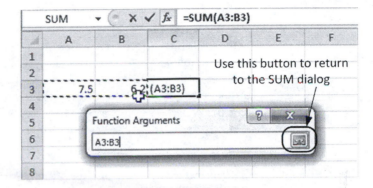

7. When you are satisfied with the selection, click on the small button at the right side of the input field in the Function Arguments dialog box. This button, called the *Return to Dialog Box button* is indicated in Figure 3.18. It is designed to look like a data entry field on a dialog box. The Return to Dialog Box button is used to jump back to the full-sized Function Arguments dialog box. When you return to the full-sized dialog box, the range you selected with the mouse will appear in the **Number1** data entry field. This is illustrated in Figure 3.19.

**Figure 3.19**
The Function Arguments dialog box after selecting the values to be summed.

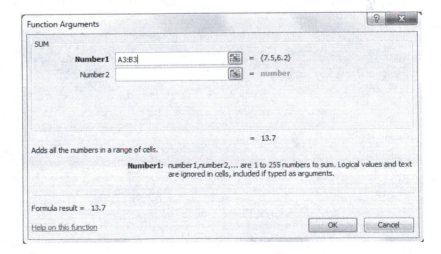

8. Click **OK** to complete the *SUM* formula in cell C3. The result is shown in Figure 3.20.

**Figure 3.20**

The result of summing the values in cells A3 and B3.

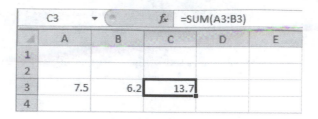

Be careful not to insert a range of cells into an expression that doesn't make sense mathematically. For example, it makes no sense to try to take the square root of a range of values, since the square root function, *SQRT*, accepts only a single value as an argument. If you attempted to send an invalid range into the *SQRT* function, such as =SQRT(A3:B3), Excel would respond by displaying the following error message in the cell containing the formula:

#VALUE

This error message is read "not a value" and indicates that you tried to pass into the *SQRT* function something other than a simple value.

### 3.4.1 Examples of Statistical Functions

The following steps will walk you through the use of two simple statistical functions that compute the mean and median of a list of numbers:

1. Enter the following seven midterm grades into the range A2:A8 as shown in Figure 3.21.

   32, 68, 93, 87, 75, 96, 82

**Figure 3.21**

Entering midterm grades and assigning a name to the cell range.

| Midterm ▾ | | *fx* 32 | | |
|---|---|---|---|---|
| | A | B | C | D | E |
| 1 | Midterm Grades | Mean Grade | Median Grade | | |
| 2 | 32 | | | | |
| 3 | 68 | | | | |
| 4 | 93 | | | | |
| 5 | 87 | | | | |
| 6 | 75 | | | | |
| 7 | 96 | | | | |
| 8 | 82 | | | | |
| 9 | | | | | |

2. Name the cell range **Midterm** by selecting the region A2:A8 and then entering the name in the Name box (illustrated in Figure 3.21).
3. Select cell B2, which will hold the mean (average) of the midterm grades and then select the **Insert Function** button from the Formula bar.

4. The Insert Function dialog box will appear. Select **Statistical** from the category list, and select **Average** from the function list, as illustrated in Figure 3.22. Click **OK**.

**Figure 3.22**
Selecting the *AVERAGE* function.

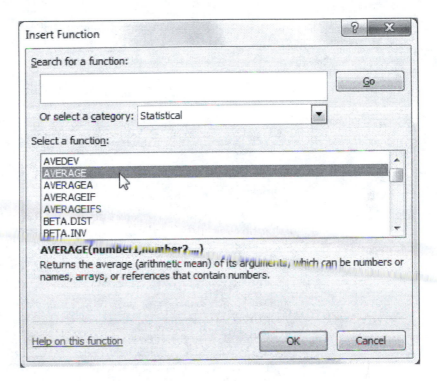

5. The Function Arguments dialog box will appear. Type **Midterm** in the box labeled **Number1**. The completed Function Arguments dialog box should resemble Figure 3.23.

**Figure 3.23**
Indicating which values should be sent to the *AVERAGE* function.

6.  Click **OK** to finish the operation. The result is shown in Figure 3.24.

**Figure 3.24**
The completed *AVERAGE* formula.

| | A | B | C | D | E |
|---|---|---|---|---|---|
| | Midterm Grades | Mean Grade | Median Grade | | |
| 1 | | | | | |
| 2 | 32 | 76.1 | | | |
| 3 | 68 | | | | |
| 4 | 93 | | | | |
| 5 | 87 | | | | |
| 6 | 75 | | | | |
| 7 | 96 | | | | |
| 8 | 82 | | | | |
| 9 | | | | | |

B2     *fx* =AVERAGE(Midterm)

**A Quicker Way . . .**

While the 6-step process of selecting the *AVERAGE function* from the Insert Function dialog box and then completing the Function Arguments dialog for the *AVERAGE* function certainly works, after a while you will probably be looking for a quicker way to enter this commonly used function. A quicker alternative is to:

1.  Enter =AVERAGE( in cell B2.
    Function names do not have to be entered in capital letters, but you do need to include the opening parenthesis to let Excel know you are entering a function name.
2.  Select the midterm grades with the mouse.
    When you select cells A2:A8, Excel will recognize the ***named cell range*** and substitute **Midterm** for the cell range in the formula.
3.  Press **Enter** to complete the formula.
    You don't even need to enter the closing parenthesis; Excel will add that automatically.

## PRACTICE!

Use Excel's **MEDIAN *function*** to calculate the median grade in cell C2. Your finished worksheet should resemble Figure 3.25.

**Figure 3.25**
Calculating mean and median grades.

C2     *fx* =MEDIAN(Midterm)

| | A | B | C | D | E |
|---|---|---|---|---|---|
| | Midterm Grades | Mean Grade | Median Grade | | |
| 1 | | | | | |
| 2 | 32 | 76.1 | 82 | | |
| 3 | 68 | | | | |
| 4 | 93 | | | | |
| 5 | 87 | | | | |
| 6 | 75 | | | | |
| 7 | 96 | | | | |
| 8 | 82 | | | | |
| 9 | | | | | |

### 3.4.2 Examples of Trigonometric Functions

Excel provides a full slate of standard trigonometric functions, including the hyperbolic varieties. In this example, we will calculate one of the angles of a right triangle from two known side lengths. The situation is illustrated in Figure 3.26.

**Figure 3.26**

A 3-4-5 triangle, used to demonstrate Excel's trigonometric functions.

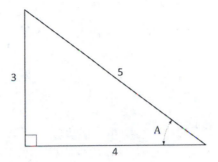

This is known as a 3-4-5 triangle, a commonly used right triangle. Since we know the lengths of all three sides, there are several ways to determine angle A:

$$\sin(A) = 3/5, \text{ so } A = \operatorname{asin}(3/5)$$

$$\cos(A) = 4/5, \text{ so } A = \operatorname{acos}(4/5)$$

$$\tan(A) = 3/4, \text{ so } A = \operatorname{atan}(3/4)$$

Any of these equations can be used. To demonstrate Excel's trigonometry functions, we will use them all.

**Note:** Excel's trigonometry functions express angles in *radians*; however, Excel provides the *DEGREES function* to easily convert radians to *degrees*.

1. Enter the known side lengths in cells B3:B5. Name cells B3, B4, and B5 "Opp," "Adj," and "Hyp," respectively. This is illustrated in Figure 3.27.
2. Use the arcsine function (*ASIN*) to compute angle A in radians (Figure 3.28).
3. Use the *DEGREES* function to convert radians to degrees (Figure 3.29).
4. Use the *ACOS function* to calculate the angle (Figure 3.30).
5. Use the *ATAN function* to calculate the angle (Figure 3.31).

**Figure 3.27**

Known data for the trigonometry example.

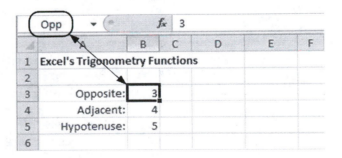

**Figure 3.28**
Using Excel's *ASIN* function to calculate an angle in radians.

| | D8 | | | $f_x$ | =ASIN(Opp/Hyp) | |
|---|---|---|---|---|---|---|
| | A | B | C | D | E | F |
| 1 | Excel's Trigonometry Functions | | | | | |
| 2 | | | | | | |
| 3 | Opposite: | 3 | | | | |
| 4 | Adjacent: | 4 | | | | |
| 5 | Hypotenuse: | 5 | | | | |
| 6 | | | | | Angle, A | |
| 7 | | | | Radians | Degrees | |
| 8 | ArcSine (ASIN) | | | 0.644 | | |
| 9 | | | | | | |

**Figure 3.29**
Converting an angle in radians to degrees.

| | E8 | | | $f_x$ | =DEGREES(D8) | |
|---|---|---|---|---|---|---|
| | A | B | C | D | E | F |
| 1 | Excel's Trigonometry Functions | | | | | |
| 2 | | | | | | |
| 3 | Opposite: | 3 | | | | |
| 4 | Adjacent: | 4 | | | | |
| 5 | Hypotenuse: | 5 | | | | |
| 6 | | | | | Angle, A | |
| 7 | | | | Radians | Degrees | |
| 8 | ArcSine (ASIN) | | | 0.644 | 36.9 | |
| 9 | | | | | | |

**Figure 3.30**
Using Excel's *ACOS* function to calculate an angle in radians.

| | D9 | | | $f_x$ | =ACOS(Adj/Hyp) | |
|---|---|---|---|---|---|---|
| | A | B | C | D | E | F |
| 1 | Excel's Trigonometry Functions | | | | | |
| 2 | | | | | | |
| 3 | Opposite: | 3 | | | | |
| 4 | Adjacent: | 4 | | | | |
| 5 | Hypotenuse: | 5 | | | | |
| 6 | | | | | Angle, A | |
| 7 | | | | Radians | Degrees | |
| 8 | ArcSine (ASIN) | | | 0.644 | 36.9 | |
| 9 | ArcCosine (ACOS) | | | 0.644 | 36.9 | |
| 10 | | | | | | |

**Figure 3.31**
Using Excel's *ATAN* function to calculate an angle in radians.

| | D10 | | | $f_x$ | =ATAN(Opp/Adj) | |
|---|---|---|---|---|---|---|
| | A | B | C | D | E | F |
| 1 | Excel's Trigonometry Functions | | | | | |
| 2 | | | | | | |
| 3 | Opposite: | 3 | | | | |
| 4 | Adjacent: | 4 | | | | |
| 5 | Hypotenuse: | 5 | | | | |
| 6 | | | | | Angle, A | |
| 7 | | | | Radians | Degrees | |
| 8 | ArcSine (ASIN) | | | 0.644 | 36.9 | |
| 9 | ArcCosine (ACOS) | | | 0.644 | 36.9 | |
| 10 | ArcTangent (ATAN) | | | 0.644 | 36.9 | |
| 11 | | | | | | |

## PRACTICE!

Identify the Excel functions that compute the following mathematical function (use the Insert Function dialog box or the Excel Help system):

1. trigonometric sine
2. inverse hyperbolic tangent
3. natural logarithm
4. base 10 logarithm
5. raise e to a power
6. convert radians to degrees
7. convert degrees or radians

*Answers:*

SIN, ATANH, LN, LOG (or LOG10), EXP, DEGREES, RADIANS

### 3.4.3 Examples of Matrix Operations

Matrices or arrays are frequently used in the formulation and solution of engineering problems. A *matrix* is a rectangular array of elements. Elements are referenced by row and column number. The grid structure of an Excel worksheet is a natural for working with matrices. Excel has a number of built-in matrix operations that are included in the **Math & Trig** category of functions. These include the following:

MDETERM(array)—returns the matrix determinant for the named array.

MINVERSE(array)—returns the inverse of the named array.

MMULT(array1, array2)—performs matrix multiplication on the two named arrays.

Some useful matrix functions are not found in the **Math & Trig** category, for example:

TRANSPOSE(array)—interchanges the rows and columns of an array.

Many other Excel functions take cell ranges as arguments and can be used to evaluate a matrix.

#### 3.4.3.1 Matrix Addition

Matrix addition is done by adding each of the corresponding cells of two matrices. The two matrices must be of the same *order*, which means that they both have the same number of rows and the same number of columns. *Matrix order* is often denoted as the number of rows by the number of columns, or (rows × columns).

As an example, let's define two matrices of order $2 \times 2$. Matrix A is stored in cells B2:C3, and that cell range has been named "A" as indicated in Figure 3.32. Matrix B is stored in cells F2:G3, and that cell range has been named "B."

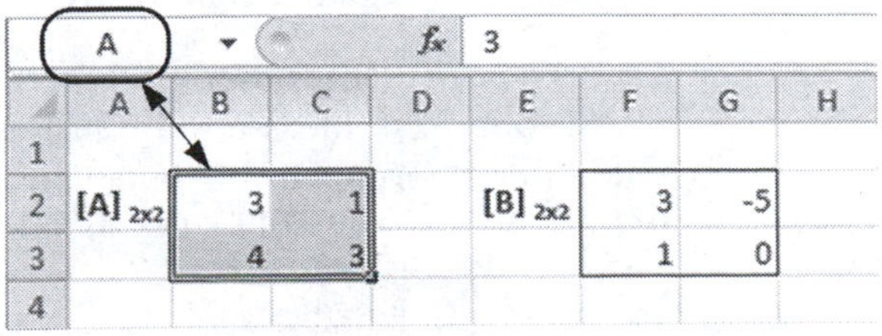

**Figure 3.32**
Defining matrices A and B.

After matrices A and B have been entered into the worksheet and named, add the matrices with these steps:

1. Select the region that will contain the result of the calculation, cells J2:K3 in this example.
2. Type the following formula (using the named matrices A and B) into the selected region.

$$= A + B$$

The result of steps 1 and 2 is shown in Figure 3.33.

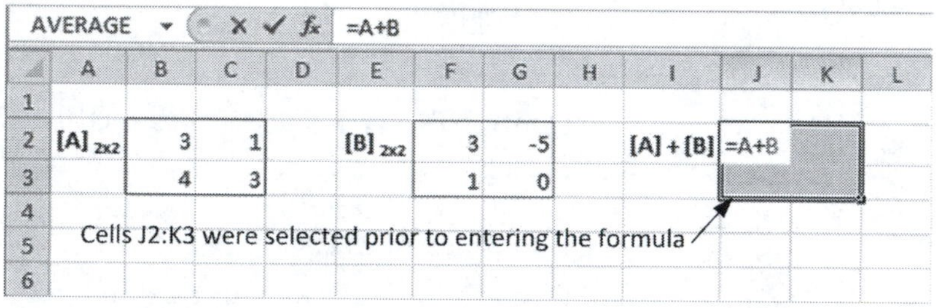

**Figure 3.33**

The not quite completed formula for adding matrices A and B.

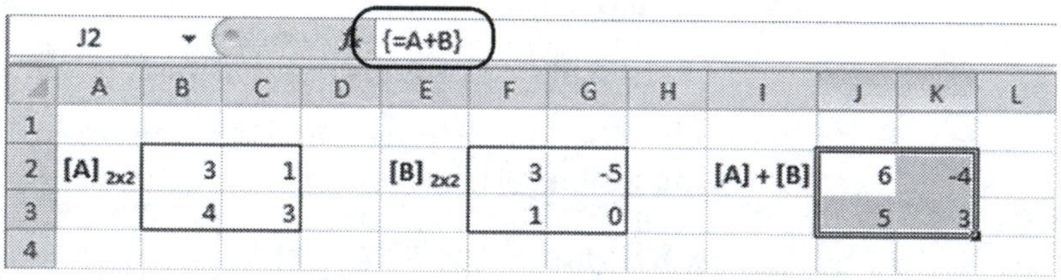

**Figure 3.34**

The sum of matrices A and B.

If you press **Enter** to complete the formula you will not get the desired result. When working with matrices you must indicate to Excel that you are asking for matrix addition. To complete an *array formula*, simultaneously press the following keys: *Ctrl + Shift + Enter*.

1. Press **Ctrl + Shift + Enter** to complete the array formula and add the matrices.

When you press **Ctrl + Shift + Enter**, Excel puts the formula in every cell in the array (selected *before* entering the formula; in step 1 of this example). The completed array formula is shown in Figure 3.34.

Notice in the Formula bar in Figure 3.34 that the formula is enclosed in curly braces. This is how Excel indicates that the formula is an array formula. Excel keeps all elements of an array together so you cannot edit or delete one cell in a cell range holding an array formula—you have to select the entire array before editing or deleting.

**Note:** You cannot type the braces to enter an array formula; the **Ctrl + Shift + Enter** key sequence must be used.

### 3.4.3.2 Matrix Transpose

As a second example, we will transpose a matrix. The *transpose* of a matrix is the matrix that is formed by interchanging the rows and columns of the original matrix. To find the transpose of matrix B, follow these steps:

1. Select the range that will hold the result.
   The selected range must be right size. Since matrix B is of order $2 \times 2$, the transpose of B will also be order $2 \times 2$. Cells F5:G6 have been selected in Figure 3.35.

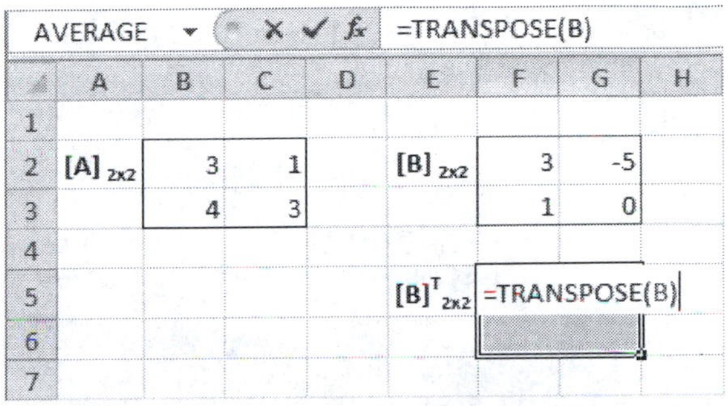

**Figure 3.35**

Entering the *TRANSPOSE* formula into cell range F5:G6.

2. Type the following formula: =TRANSPOSE(B).
3. Press **Ctrl + Shift + Enter** to complete the array formula.
4. Place appropriate borders and labels on your worksheet. The results should resemble Figure 3.36.

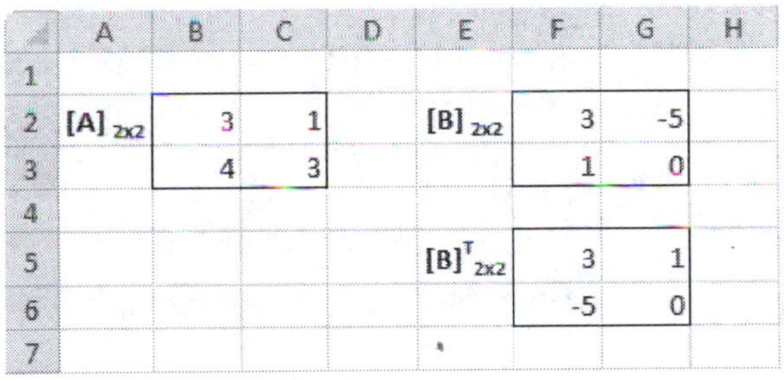

**Figure 3.36**

Computing the transpose of matrix B.

### 3.4.3.3 Multiplying Matrices

Matrix multiplication is defined as follows: If A = $[a_{ij}]$ is an $m \times n$ matrix and B = $[b_{ij}]$ is an $n \times p$ matrix, then the product AB = C = $[c_{ij}]$ is an $m \times p$ matrix defined by

$$c_{ij} = \sum_{k=1}^{n} a_{ik} b_{kj}, \quad i = 1, 2, \ldots, m, \quad j = 1, 2, \ldots, p.$$

Using this equation, the product of A and B is calculated as

$$AB = \begin{bmatrix} 3 & 1 \\ 4 & 3 \end{bmatrix} \begin{bmatrix} 3 & -5 \\ 1 & 0 \end{bmatrix}$$

$$= \begin{bmatrix} (3 \cdot 3) + (1 \cdot 1) & (3 \cdot -5) + (1 \cdot 0) \\ (4 \cdot 3) + (3 \cdot 1) & (4 \cdot -5) + (3 \cdot 0) \end{bmatrix} = \begin{bmatrix} 10 & -15 \\ 15 & -20 \end{bmatrix}.$$

Excel has a built-in matrix multiplication function named ***MMULT***. In Figure 3.37, we have used this function to verify the preceding results. You can practice using this function even if you have not yet studied matrix multiplication. Remember that matrix functions must be entered by using **Ctrl + Shift + Enter**.

**Figure 3.37**
Using *MMULT* to multiply two matrices.

| | J2 | | | $f_x$ {=MMULT(A,B)} | | | | | | | |
|---|---|---|---|---|---|---|---|---|---|---|---|
| | A | B | C | D | E | F | G | H | I | J | K | L |
| 1 | | | | | | | | | | | | |
| 2 | [A]$_{2x2}$ | 3 | 1 | | [B]$_{2x2}$ | 3 | -5 | | [AB] | 10 | -15 | |
| 3 | | 4 | 3 | | | 1 | 0 | | | 15 | -20 | |
| 4 | | | | | | | | | | | | |

### 3.4.4 Examples of Financial Functions

Suppose that you want to predict the future value of an investment at two different interest rates. The equation relating the *present value*, P, to the *future value*, F, is

$$F = P[1 + i]^N$$

The present value is the amount you invest, and the future value is what the investment will be worth at N years later. The i in the equation is the annual interest rate.

> **Note:** Officially, N is the number of compounding periods, and i is the interest rate per period—the period doesn't have to be a year. Monthly compounding is common. For example, for a five-year investment at 12% annual percentage rate (APR) with monthly compounding, $N = 5 \times 12 = 60$ months and $i = 12\%/12 = 1\%$ per month.

Excel provides a set of financial functions that can solve this equation directly. In this example, John wants to know how much a $10,000 investment would be worth in 40 years if it were invested at 6% per year (assuming annual compounding). He also wants to compare that result with the future value if the funds were invested at 11% per year.

**Case A: 6% Interest (Compounded Annually)**

From the equation, we can see that we need to know P, i, and N in order to solve for F. So, first we enter the known values into a worksheet, as shown in Figure 3.38.

**Figure 3.38**
Entering known values before calculating the future value of an investment.

| | A | B | C | D | E |
|---|---|---|---|---|---|
| 1 | Calculating the Future Value of an Investment | | | | |
| 2 | | | | | |
| 3 | Case A: 6% Interest | | | | |
| 4 | | | | | |
| 5 | P: | -$10,000 | expense | | |
| 6 | I: | 6% | per year | | |
| 7 | N: | 40 | years | | |
| 8 | | | | | |

Notice that the present value was entered as a negative number. There is a standard convention that Excel uses to distinguish incomes and expenses. Expenses are indicated as negative, and incomes are indicated as positive. Since you must put money out of your pocket into the investment, that is an expense and is entered as a negative value.

To calculate the future value, follow these steps:

1. Click in cell B9, and then click the **Insert Function** button on the Formula Bar.
2. Select the *FV* (future value) function from the Insert Function dialog box (see Figure 3.39). The Function Arguments dialog box for the FV function will be displayed (Figure 3.40).
3. Fill in the data entry fields for function FV's arguments. The completed dialog box is shown in Figure 3.40.
4. Click **OK** to complete the calculation. The result is shown in Figure 3.41.

**Figure 3.39**

Selecting the *FV* function from the Financial category.

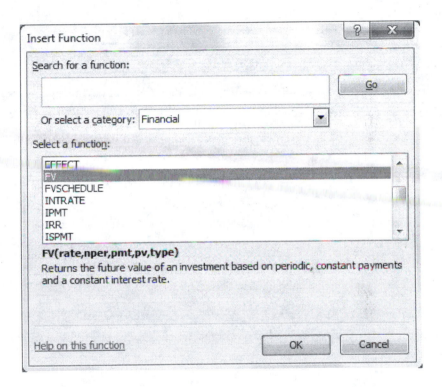

**Figure 3.40**

Specifying the function arguments for the *FV* function.

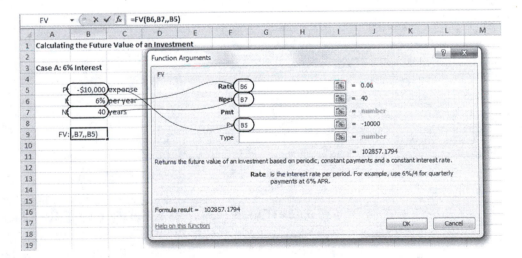

**Figure 3.41**

The calculated future value at 6% interest.

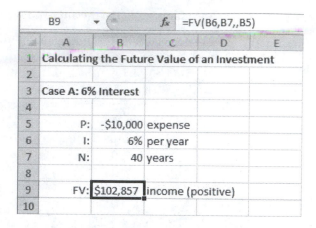

Notice that the Pmt (annual payment) field was left blank in Figure 3.40. This is necessary because there was no mention of any annual payments, just an initial investment of $10,000. This shows up in the formula in cell B9 as an unused argument (notice the two adjacent commas in the formula shown in Figure 3.41). You could also put a value of zero in the Pmt field.

**Case B: 11% Interest (Compounded Annually)**

Now, repeat these steps, using an 11% annual rate. The result is shown in Figure 3.42.

**Figure 3.42**

Calculating the future value at 11% interest.

| B17 | | $f_x$ | =FV(B14,B15,,B13) | |
|---|---|---|---|---|
| | A | B | C | D | E |
| 1 | Calculating the Future Value of an Investment | | | | |
| 2 | | | | | |
| 3 | Case A: 6% Interest | | | | |
| 4 | | | | | |
| 5 | P: | -$10,000 | expense | | |
| 6 | I: | 6% | per year | | |
| 7 | N: | 40 | years | | |
| 8 | | | | | |
| 9 | FV: | $102,857 | income (positive) | | |
| 10 | | | | | |
| 11 | Case B: 11% Interest | | | | |
| 12 | | | | | |
| 13 | P: | -$10,000 | expense | | |
| 14 | I: | 11% | per year | | |
| 15 | N: | 40 | years | | |
| 16 | | | | | |
| 17 | FV: | $650,009 | income (positive) | | |
| 18 | | | | | |

## 3.5 ABSOLUTE AND RELATIVE CELL REFERENCES

Formulas may be copied from one location to another in a worksheet. Usually, you will want the cells that are referenced in the formula to follow the formula—this is called *relative referencing*.

For example, you may want to copy a formula that adds a row of numbers. Look at Figure 3.43. The formula in cell E1 adds the elements in cells A1 to C1. If you were to copy the formula in cell E1 to cell E2, what result would you want? Should the new formula compute the sum using values in row 1 or row 2?

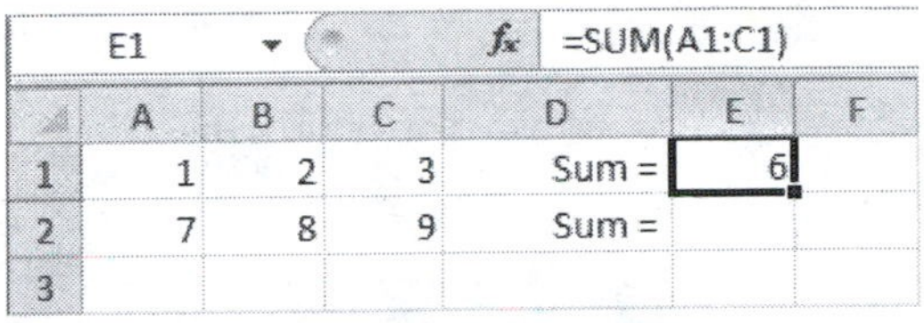

**Figure 3.43**

Example of a formula with relative referencing.

| E1 | | | | $f_x$ | =SUM(A1:C1) | |
|---|---|---|---|---|---|---|
| | A | B | C | D | E | F |
| 1 | 1 | 2 | 3 | Sum = | 6 | |
| 2 | 7 | 8 | 9 | Sum = | | |
| 3 | | | | | | |

If relative referencing is used, the copied formula in cell E2 will sum values in row 2 (cells A2 to C2). Try the following test:

1. Create the worksheet shown in Figure 3.43.
2. Copy the formula in cell E1 to cell E2 (you can use the **Fill Handle** to do this easily).

The result is shown in Figure 3.44. Note that the formula in cell E2 adds values on row 2, not row 1. The cell references have followed the formula (relative referencing).

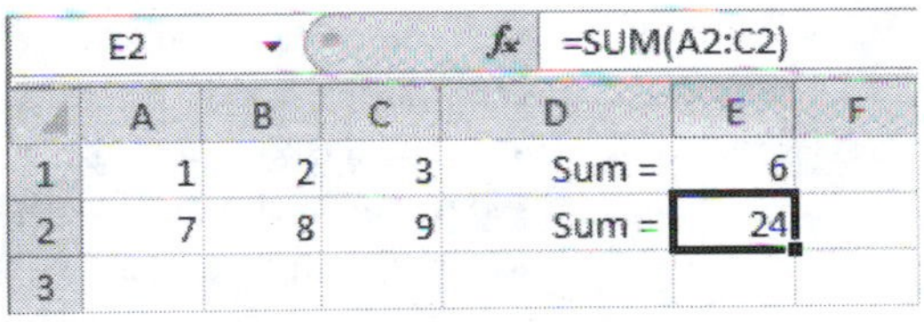

**Figure 3.44**

Copying a formula that uses relative referencing.

| E2 | | | | $f_x$ | =SUM(A2:C2) | |
|---|---|---|---|---|---|---|
| | A | B | C | D | E | F |
| 1 | 1 | 2 | 3 | Sum = | 6 | |
| 2 | 7 | 8 | 9 | Sum = | 24 | |
| 3 | | | | | | |

There are times when you may want to copy a formula and *not* have the references follow the formula. This is called **absolute referencing**. A cell or range is denoted as an *absolute reference* by the placement of a dollar sign ($) in front of the row or column to be locked.

Look at Figure 3.45. This worksheet is the same as the worksheet in Figure 3.43, except that the formula uses absolute references, indicated by the dollar signs in the cell references.

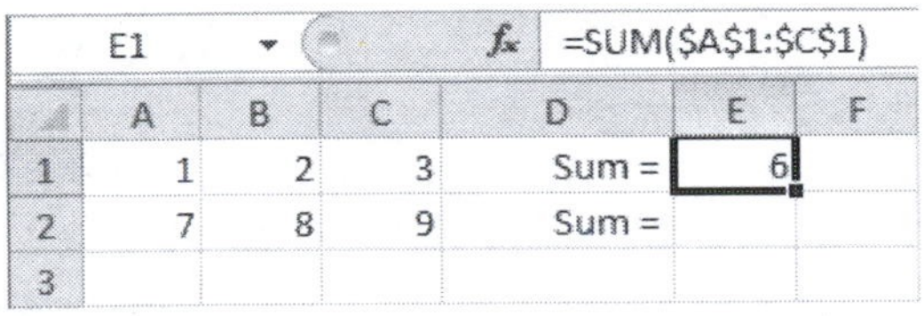

**Figure 3.45**

Example of a formula with absolute cell references ($).

| E1 | | | | $f_x$ | =SUM($A$1:$C$1) | |
|---|---|---|---|---|---|---|
| | A | B | C | D | E | F |
| 1 | 1 | 2 | 3 | Sum = | 6 | |
| 2 | 7 | 8 | 9 | Sum = | | |
| 3 | | | | | | |

When cell E1 is copied to cell E2 the results (Figure 3.46) are quite different with absolute cell referencing than with relative referencing: the formula in cell E2 sums values in row 1, not row 2! The references have not followed the formula.

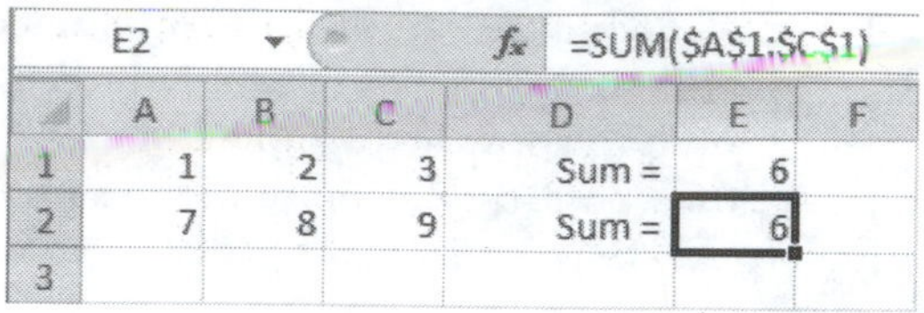

**Figure 3.46**
The result of copying the formula in cell E1 to cell E2.

| | | | fx | =SUM($A$1:$C$1) |
| --- | --- | --- | --- | --- |

| | A | B | C | D | E | F |
| --- | --- | --- | --- | --- | --- | --- |
| 1 | 1 | 2 | 3 | Sum = | 6 | |
| 2 | 7 | 8 | 9 | Sum = | 6 | |
| 3 | | | | | | |

Each dollar sign (absolute reference symbol) acts like a lock on the cell reference.

- $A1 tells Excel not to change the A when the cell is copied (but the row reference can change).
- A$1 tells Excel not to change the 1 when the cell is copied (but the column reference can change).
- $A$1 tells Excel not to change either the A or the 1 when the cell is copied.

## PRACTICE!

There are times when you may want to make relative references to some cells and absolute references to other cells. For example, a formula may use a constant and several variables. You can use absolute referencing for the constant and relative referencing for the variables.

In the next example, we want to compute the areas of several circles with various radii. The formula uses the constant $\pi$ (pi). When we copy the formula, we want the reference to $\pi$ to remain absolute, but the reference to the radius to be relative. To practice with relative and absolute cell addresses, complete the following steps:

1. Create the worksheet shown in Figure 3.47. Notice that the formula for area in cell B6 uses absolute referencing for cell $B$3 (the constant, $\pi$), and relative referencing for cell A6 (the radius value).

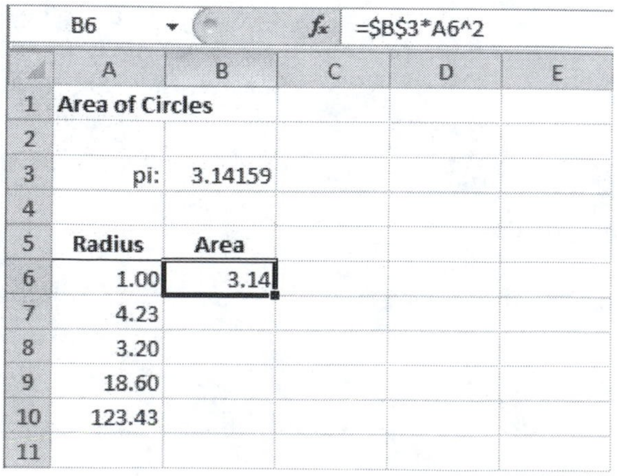

**Figure 3.47**
Using both absolute and relative cell references.

| B6 | | | fx | =$B$3*A6^2 |
| --- | --- | --- | --- | --- |

| | A | B | C | D | E |
| --- | --- | --- | --- | --- | --- |
| 1 | Area of Circles | | | | |
| 2 | | | | | |
| 3 | pi: | 3.14159 | | | |
| 4 | | | | | |
| 5 | Radius | Area | | | |
| 6 | 1.00 | 3.14 | | | |
| 7 | 4.23 | | | | |
| 8 | 3.20 | | | | |
| 9 | 18.60 | | | | |
| 10 | 123.43 | | | | |
| 11 | | | | | |

2. Copy the formula in cell B6 by dragging the Fill Handle through cells B7:B10.

The result is shown in Figure 3.48. Because the formula in cell B6 used relative referencing for the radius value (A6), when the formula was copied the radius references were updated to use the values in cells A7 to A10. But because the reference to the cell containing the value of π used absolute referencing ($B$3), all of the copied formulas in cells B7 to B10 still refer to the value of π stored in cell B3.

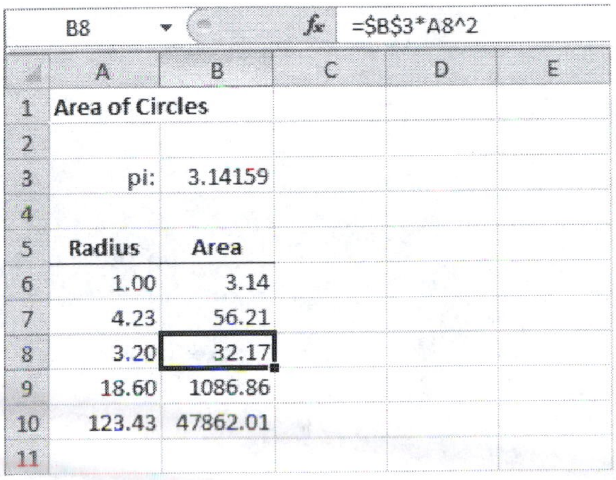

**Figure 3.48**
The result of copying the formula in cell B6 to B7:B10.

| | B8 | ▼ | fx | =$B$3*A8^2 | |
|---|---|---|---|---|---|
| | **A** | **B** | **C** | **D** | **E** |
| 1 | Area of Circles | | | | |
| 2 | | | | | |
| 3 | pi: | 3.14159 | | | |
| 4 | | | | | |
| 5 | **Radius** | **Area** | | | |
| 6 | 1.00 | 3.14 | | | |
| 7 | 4.23 | 56.21 | | | |
| 8 | 3.20 | 32.17 | | | |
| 9 | 18.60 | 1086.86 | | | |
| 10 | 123.43 | 47862.01 | | | |
| 11 | | | | | |

## PRACTICE!

An example of a slightly more complex use of formulas is the solution of *quadratic equations*. You may recall from high school algebra that if a quadratic equation is expressed in the form

$$ax^2 + bx + c = 0,$$

then the solutions for *x* are as follows:

$$x = -b \pm \frac{\sqrt{b^2 - 4ac}}{2a}, \ (2a \neq 0)$$

Note that there are two solutions because of the ± in the equation. Since Excel does not directly recognize imaginary numbers, we must make the further restriction that

$$b^2 - 4ac \geq 0.$$

Practice working with Excel formulas by creating a worksheet that resembles Figure 3.49. Store coefficients in columns A, B, and C and enter the formulas for the two solutions in cells E5 and F5, respectively. The Excel formula for the first solution (in cell E5) is

$$= (-B5 + SQRT(B5^2 - 4*A5*C5))/(2*A5)$$

The formula for the second solution (in cell F5) is

$$= (-B5 - SQRT(B5^2 - 4*A5*C5))/(2*A5)$$

Verify that the constraints are being met by checking that $2a \neq 0$ (cell H5) and $b^2 - 4ac \geq 0$ (cell I5).

**Figure 3.49**
Solving quadratic equations (partial).

| E5 | | | | $fx$ =(-B5+SQRT(B5^2-4*A5*C5))/(2*A5) | | | | | |
|---|---|---|---|---|---|---|---|---|---|
| | A | B | C | D | E | F | G | H | I | J |

**Solving Quadratic Equations**

| | a | b | c | Root 1 | Root 2 | | 2a | b² - 4ac |
|---|---|---|---|---|---|---|---|---|
| | | Coefficients | | | Solutions | | | Checks | |
| 5 | 1 | 2 | 1 | -1.00 | -1.00 | | 2.00 | 0.00 |
| 6 | 1 | 16 | 1 | | | | | |
| 7 | -4 | -8 | 24 | | | | | |
| 8 | 0 | 18 | 24 | | | | | |
| 9 | -4 | -8 | 4 | | | | | |

Next, use the Fill Handle to copy the formulas in cells E5, F5, H5, and I5 down through rows 6 to 9. Your results should resemble Figure 3.50.

**Figure 3.50**
Solving quadratic equations (complete).

| E6 | | | | $fx$ =(-B6+SQRT(B6^2-4*A6*C6))/(2*A6) | | | | | |
|---|---|---|---|---|---|---|---|---|---|
| | A | B | C | D | E | F | G | H | I | J |

**Solving Quadratic Equations**

| | | Coefficients | | | Solutions | | | Checks | |
|---|---|---|---|---|---|---|---|---|---|
| | a | b | c | Root 1 | Root 2 | | 2a | b² - 4ac |
| 5 | 1 | 2 | 1 | -1.00 | -1.00 | | 2.00 | 0.00 |
| 6 | 1 | 16 | 1 | -0.06 | -15.94 | | 2.00 | 252.00 |
| 7 | -4 | -8 | 24 | -3.65 | 1.65 | | -8.00 | 448.00 |
| 8 | 0 | 18 | 24 | #DIV/0! | #DIV/0! | | 0.00 | 324.00 |
| 9 | -4 | -8 | 4 | -2.41 | 0.41 | | -8.00 | 128.00 |

Notice how the row numbers in the formulas change as you are using relative references. Also, notice that an error is displayed for the solutions on row 8. The check calculation in cell H8 shows why; we have violated the constraint that $2a \neq 0$.

## 3.6 EXCEL ERROR MESSAGES

The formulas that we have presented so far are relatively simple. If you make an error when typing one of the example formulas, the location of the error is relatively easy to spot. As you begin to develop more complex formulas, locating and debugging errors becomes a more difficult problem.

When a syntax error occurs in a formula, Excel will attempt to immediately catch the error and then display an error box that explains the error. However, formulas can be syntactically correct, but still produce errors when the formula is executed. If an expression cannot be evaluated, then Excel will denote the error by placing one of eight *error messages* in the target cell. These error messages are listed in Table 3.2.

## 3.7 DEBUGGING EXCEL WORKSHEETS

When things go wrong and your worksheet is calculating values that can't be right, you have to find the error(s). This is called *debugging* the worksheet. With small worksheets you can select each cell and visually check to see if the formula is correct.

**Table 3.2 Excel Error Messages**

| Message | Description |
| --- | --- |
| ###### | The value is too wide to fit in the cell, or an attempt was made to display a negative date or time. |
| #VALUE | The wrong type of argument was used in a formula. This will occur, for example, if text were entered when an array argument was expected. |
| #DIV/0 | An attempt was made to divide by zero in a formula. See the quadratic equation example in Figure 3.50 for an illustration. |
| #NAME | A name used in a formula is not recognized. Usually, the function or defined name was misspelled. Note that named ranges or functions may not contain spaces. |
| #REF | A referenced cell is not valid. This usually occurs when a cell is referenced in a formula and that cell is then deleted. It also occurs if an attempt is made to paste a cell over a referenced cell. |
| #NUM | The expression produces a numeric value that is out of range or invalid. Examples are extremely small, large, or imaginary numbers. To see this error, try this formula: =SQRT(-1). |
| #NULL | An attempt was made to reference the intersection of two areas that don't intersect. This usually occurs when a space is inadvertently placed between two arguments, instead of a comma or colon, as in =SUM(C2 D3). |

But when worksheets get really large, checking every cell for an error is difficult, if not impossible. Fortunately, Excel provides some debugging tools that can help.

The worksheet shown in Figure 3.51 is a long-handed way of calculating the mean and standard deviation of a set of grades. Excel has a built-in function for calculating standard deviation (***STDEV***), but we need a worksheet that contains an error to show some debugging options. The worksheet in Figure 3.51 contains an error—finding the error can be a challenge.

The standard deviation can be determined using the equation

$$\text{Std. Dev.} = \sqrt{\frac{\sum_{i=1}^{N}(x_i - \bar{x})^2}{N-1}}$$

where

$$\text{Deviation} = x_i - \bar{x}$$

**Figure 3.51**
A worksheet containing an error.

| | A | B | C | D | E | F | G |
| --- | --- | --- | --- | --- | --- | --- | --- |
| 1 | Grade Calculations | | | | | | |
| 2 | | | | | | | |
| 3 | Grades | Deviation | Dev$^2$ | | | Results | |
| 4 | 32 | 0.71 | 0.51 | | | | |
| 5 | 14 | -17.29 | 298.80 | | Mean: | 31.29 | |
| 6 | 52 | 20.71 | 429.08 | | Sum(Dev$^2$): | 1003.91 | |
| 7 | 26 | -5.29 | 27.94 | | Std. Dev.: | 11.93 | |
| 8 | 18 | -13.29 | 176.51 | | | | |
| 9 | 45 | 8.40 | 70.56 | | | | |
| 10 | 32 | 0.71 | 0.51 | | | | |
| 11 | | | | | | | |

### 3.7.1 Highlighting Formulas

A common worksheet error is the accidental replacement of a formula with a constant. A quick way to check for this is to have Excel highlight all cells that contain formulas. In Excel 2007 and 2010 use Ribbon options **Home tab → Editing group → Find & Select** drop-down menu → **Formulas option**. [Excel 2003: Edit → Go To → Select "Special" on the Go To dialog → Select "Formulas" on the Go To Special dialog.]

Excel will highlight all cells containing formulas, as illustrated in Figure 3.52. With the formulas highlighted it quickly becomes apparent that there is something wrong in cell B9; it doesn't contain a formula. Someone has typed 8.4 into cell B9, overwriting the formula.

**Figure 3.52**
The grade calculation worksheet with all formulas highlighted.

| | A | B | C | D | E | F | G |
|---|---|---|---|---|---|---|---|
| 1 | Grade Calculations | | | | | | |
| 2 | | | | | | | |
| 3 | Grades | Deviation | Dev$^2$ | | | Results | |
| 4 | 32 | 0.71 | 0.51 | | | | |
| 5 | 14 | -17.29 | 298.80 | | Mean: | 31.29 | |
| 6 | 52 | 20.71 | 429.08 | | Sum(Dev$^2$): | 1003.91 | |
| 7 | 26 | -5.29 | 27.94 | | Std. Dev.: | 11.93 | |
| 8 | 18 | -13.29 | 176.51 | | | | |
| 9 | 45 | 8.40 | 70.56 | | Cell B9 does not | | |
| 10 | 32 | 0.71 | 0.51 | | contain a formula | | |
| 11 | | | | | | | |

Another option to check for this type of error is to have Excel display all of the formulas in the worksheet. In Excel 2007 and 2010 this is done by toggling the **Show Formulas** button on the Ribbon (**Formulas tab → Formula Auditing group → Show Formulas button**). When formulas are displayed (Figure 3.53), the error in cell B9 is again fairly easy to spot.

**Figure 3.53**
Toggling the display of formulas to search for an error.

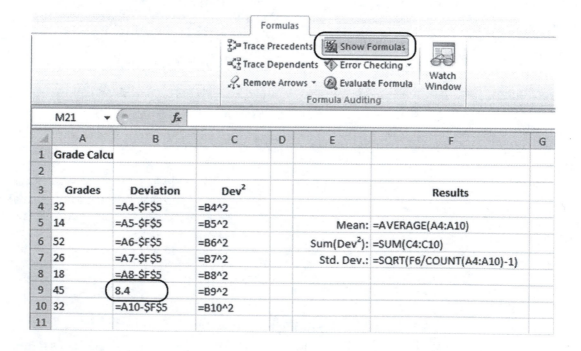

### 3.7.2 Tracing Dependents

A *dependent* is a value that is calculated after another calculation. In the grade calculations worksheet, you cannot calculate Deviation values until the Mean has been calculated, so the Deviation values are dependents of the Mean calculation.

You can have Excel display all of the dependents for any formula on the worksheet. In Excel 2007 and 2010 this is done by selecting the formula you want to investigate, and then clicking the **Trace Dependents** button on the Ribbon (**Formulas tab → Formula Auditing group → Trace Dependents button**). This is illustrated in Figure 3.54. To clear the arrows from the display, click the **Remove Arrows** button on the Ribbon's Formulas tab.

**Figure 3.54**
Using Trace Dependents to search for errors.

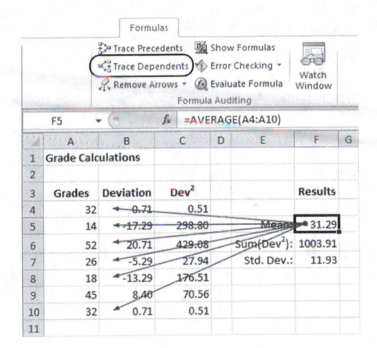

### 3.7.3 Tracing Precedents

A *precedent* is something that occurs before something else. In the grade worksheet, the Deviation values are calculated before the $Dev^2$ values, so the cells containing Deviation values are precedents for the cells containing the $Dev^2$ values. You can have Excel display all of the precedents for a particular formula by selecting the formula, then clicking the **Trace Precedents** button on the Ribbon's Formulas tab (**Formulas tab → Formula Auditing group → Trace Precedents button**). This is illustrated in Figure 3.55. To clear the arrows from the display, click the **Remove Arrows** button on the Ribbon's Formulas tab.

In Figure 3.55 we see that the calculation of the standard deviation depends on the value in cell F6 (the sum of the squared deviations) and all of the individual grades. It may seem odd that the individual grades are needed for this calculation, but Excel's *COUNT* function was used in the formula to determine the number of grades in the data set, and the *COUNT* function needed all of the grades to perform the count.

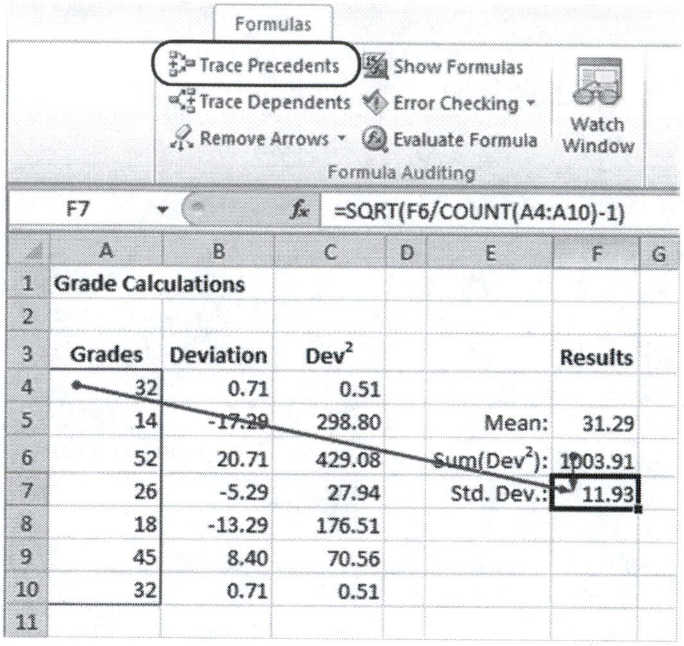

**Figure 3.55**

Checking the precedents of the standard deviation formula.

### 3.7.3.1 Column (or Row) Differences

Excel looks for unusual changes in row and column patterns and tries to let you know there may be a problem in an unobtrusive way. In Figure 3.56, the formula in cell C7 was entered as $= B7\char94 3$ instead of $= B7\char94 2$ which can happen with a simple typo while entering the formula. Excel noticed a change in pattern in that cell compared with its neighbors and marked the corner of the questionable cell with a colored triangle (indicated in Figure 3.56). These pattern changes may or may not represent errors, but they are usually worth checking out.

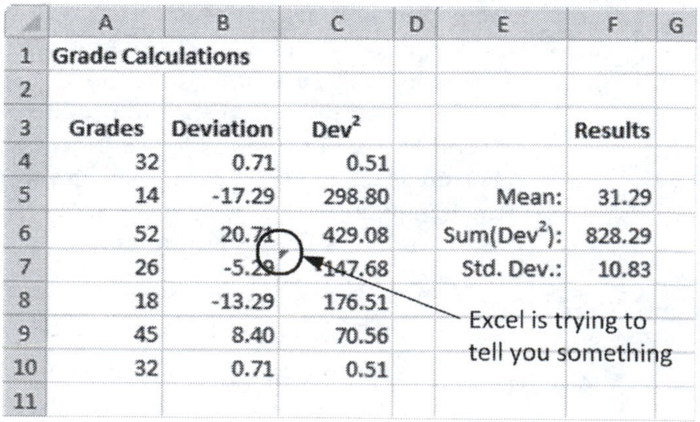

**Figure 3.56**

Excel marks the corner of cells when patterns seem to have changed.

If you select the questionable cell and move your mouse over the warning indicator, a description of the potential error will be displayed (see Figure 3.57).

**Figure 3.57**
Move the mouse over the warning icon to see a description of the potential error.

| | A | B | C | D | E | F | G | H |
|---|---|---|---|---|---|---|---|---|
| 1 | Grade Calculations | | | | | | | |
| 2 | | | | | | | | |
| 3 | Grades | Deviation | Dev$^2$ | | | Results | | |
| 4 | 32 | 0.71 | 0.51 | | | | | |
| 5 | 14 | -17.29 | 298.80 | | Mean: | 31.29 | | |
| 6 | 52 | 20.71 | 429.08 | | Sum(Dev$^2$): | 828.29 | | |
| 7 | 26 | ⚠ ▾ | -147.68 | | Std. Dev.: | 10.83 | | |
| 8 | 18 | -13. | 176.51 | | | | | |
| 9 | | The formula in this cell differs from the formulas in this area of the spreadsheet. | | | | | | |
| 10 | 32 | 0.71 | 0.51 | | | | | |
| 11 | | | | | | | | |

Excel's ability to watch out for row and column differences is a nice feature, but it has a couple of limitations:

- A value typed in over a formula (as in cell B9) is not detected.
- A change in pattern in the first or last cell of a column (or row) is not detected.

Several of the debugging methods described here helped identify an error in cell B9 (constant typed in over a formula) and a second error (typo in a formula) inadvertently created while trying to find the first error. With those errors fixed, we can finally calculate the standard deviation of the grade data correctly (Figure 3.58).

**Figure 3.58**
Corrected grade calculations worksheet.

| | A | B | C | D | E | F | G |
|---|---|---|---|---|---|---|---|
| 1 | Grade Calculations | | | | | | |
| 2 | | | | | | | |
| 3 | Grades | Deviation | Dev$^2$ | | | Results | |
| 4 | 32 | 0.71 | 0.51 | | | | |
| 5 | 14 | -17.29 | 298.80 | | Mean: | 31.29 | |
| 6 | 52 | 20.71 | 429.08 | | Sum(Dev$^2$): | 1121.43 | |
| 7 | 26 | -5.29 | 27.94 | | Std. Dev.: | 12.62 | |
| 8 | 18 | -13.29 | 176.51 | | | | |
| 9 | 45 | 13.71 | 188.08 | | | | |
| 10 | 32 | 0.71 | 0.51 | | | | |
| 11 | | | | | | | |

The debugging tools prove their worth as the worksheet gets larger and more complex. Consider trying to find the same error without the debugging tool if column A contained 3,000 grades instead of 7 grades!

## 3.8 USING MACROS TO AUTOMATE COMPUTATIONS

A *macro* is a stored collection of commands. If you repeat the same set of commands several times, then using a macro can be a convenient, time-saving feature.

A macro is stored internally in a Visual Basic module. Visual Basic is a programming language, and it is not within the scope of this text to teach you the Visual Basic language, and Excel allows you to record and execute macros without knowing Visual Basic. But because there is a programming language built into Excel (and other Microsoft Office programs), there is a risk of contracting a computer virus (called a *macro virus*) that can be written in Visual basic.

With Excel 2007, Microsoft took a big step toward minimizing the spread of macro viruses by using a new file name extension for most Excel workbooks, .xlsx. That final "x" is an indicator that macros cannot be stored in the workbook, and therefore the workbook cannot transmit a macro virus. You can write and use macros in any Excel workbook, but you can only save the macros if you use the new .xlsm file extension. This extension is available as part of the Save As procedure.

### 3.8.1 Recording a Macro

In Excel you can record a macro, or write a macro program. In the next sections, you will be guided through recording and running a macro. Then you will be shown how to view the Visual Basic code that contains the macro commands. If you were to learn Visual Basic, then you could edit the code directly or write your own macros in the Visual Basic language.

Before recording a macro, it is wise to carefully plan the steps that you will be taking. When you are in recording mode, everything that you type is recorded—mistakes and all. In the next example, the major steps for computing several statistics for a set of data are listed. It is assumed that you are familiar with the use of Excel's built-in mathematical functions.

To prepare for recording the statistics macro, create the worksheet shown in Figure 3.59.

**Figure 3.59**

Preparing to record a macro.

| | A | B | C | D | E | F |
|---|---|---|---|---|---|---|
| 1 | Statistical Calculations | | | | | |
| 2 | | | | | | |
| 3 | | Data | | | | |
| 4 | | 10 | | | | |
| 5 | | 12 | | | | |
| 6 | | 45 | | | | |
| 7 | | 32 | | | | |
| 8 | | 23 | | | | |
| 9 | | 23 | | | | |
| 10 | | 76 | | | | |
| 11 | | 21 | | | | |
| 12 | | 32 | | | | |
| 13 | | 21 | | | | |
| 14 | | | | | | |

Next, decide if you want to record the macro using *relative references,* or *absolute references.* For example, we are going to calculate the mean value of the data set and put the result in cell E4, which is three cells to the right of the top of the column of data values. We have to tell Excel whether we want the result "three cells to the right of the top of the column of data values" (relative reference), or "in cell E4" (absolute reference). It is usually more useful to use relative referencing in macros.

To activate relative referencing, use the Ribbon's View tab with the following options: **View tab → Macros group → Macros drop-down menu → Click the Use Relative References toggle button.** The **Relative References** toggle button is highlighted (see Figure 3.60) when relative references are active.

**Figure 3.60**

The icon on the Use Relative References toggle button has a colored background when selected.

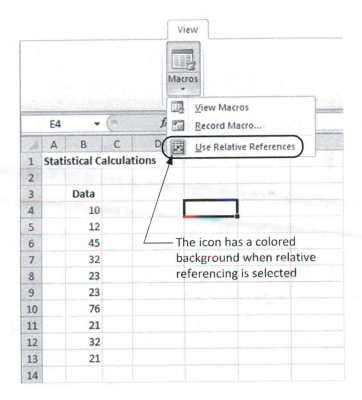

One last step before recording the macro: choosing the cell that should be selected when macro recording begins. There are a few commonly used choices:

1. Select the cell at the top of the column of data. The advantage is that it is fairly easy to select the entire data set from this starting position.
2. Select the entire data set. This is commonly used if the macro will perform one calculation on the selected data (not so in this example).
3. Select the location where the result will be placed.

We will choose the first option in this example, although the other options could be used as well. Select cell B4 before recording the macro.

To record the sample macro, perform the following steps:

1. Turn on macro recording using the following Ribbon options: **View tab → Macros group → Macros drop-down menu → Record Macro . . . button**. [Excel 2003: Tools → Macro → Record New Macro.] The Record Macro dialog box will appear, as shown in Figure 3.61.

**Figure 3.61**
The Record Macro dialog box.

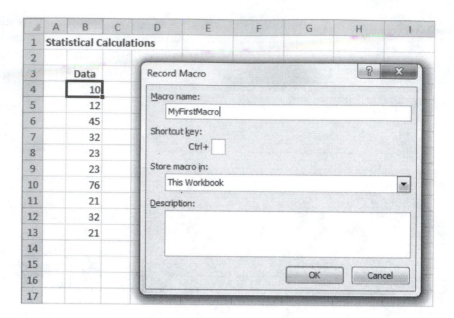

2. Give your macro a name—this example uses MyFirstMacro. This dialog box also allows you to assign a shortcut key and choose where to store the macro. When you are ready to record the macro, click **OK**.

**Note:** The "This Workbook" option is the default storage location, but your macro will not be stored with the workbook unless you use a macro-enabled workbook with the .xlsm file extension. If you try to save your workbook with the .xslx file extension after recording a macro, Excel will warn you that the macro cannot be stored.

CAUTION! Everything you now type will be recorded in the macro until you stop the recording process! To assist you in stopping the recording process, a Stop button is placed in the Status bar at the bottom of the Excel window, as indicated in Figure 3.62. [Excel 2003: The Stop button floats in a small dialog box.]

**Figure 3.62**
A Stop button appears in the Status bar while a macro is being recorded.

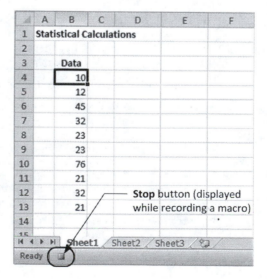

Stop button (displayed while recording a macro)

3. Enter the text "Results" in cell E3, then center and bold the font.
4. Enter the text "Mean:" in cell D4, right justified.
5. Enter the formula =AVERAGE(B4:B13) in cell E4.

At this point your worksheet should resemble Figure 3.63.

**Figure 3.63**
The worksheet after the first set of results.

| | A | B | C | D | E | F |
|---|---|---|---|---|---|---|
| 1 | Statistical Calculations | | | | | |
| 2 | | | | | | |
| 3 | | Data | | | Results | |
| 4 | | 10 | | Mean: | 29.5 | |
| 5 | | 12 | | | | |
| 6 | | 45 | | | | |
| 7 | | 32 | | | | |
| 8 | | 23 | | | | |
| 9 | | 23 | | | | |
| 10 | | 76 | | | | |
| 11 | | 21 | | | | |
| 12 | | 32 | | | | |
| 13 | | 21 | | | | |
| 14 | | | | | | |

6. Enter the text "Median:" in cell D5, right justified.
7. Enter the formula =MEDIAN(B4:B13) in cell E5.
8. Enter the text "Std. Dev.:" in cell D6, right justified.
9. Enter the formula =STDEV(B4:B13) in cell E6.
10. Check the formatting of the calculated values, adjusting the number of displayed digits as needed.
11. Press the **Stop** button to stop recording the macro. Congratulations—you have recorded a macro! Your worksheet should resemble Figure 3.64.

**Figure 3.64**
The worksheet after recording the macro.

| | A | B | C | D | E | F |
|---|---|---|---|---|---|---|
| 1 | Statistical Calculations | | | | | |
| 2 | | | | | | |
| 3 | | Data | | | Results | |
| 4 | | 10 | | Mean: | 29.5 | |
| 5 | | 12 | | Median: | 23 | |
| 6 | | 45 | | Std. Dev.: | 19.24 | |
| 7 | | 32 | | | | |
| 8 | | 23 | | | | |
| 9 | | 23 | | | | |
| 10 | | 76 | | | | |
| 11 | | 21 | | | | |
| 12 | | 32 | | | | |
| 13 | | 21 | | | | |
| 14 | | | | | | |

### 3.8.2 Running a Macro

You can now retrieve and reuse your recorded macro whenever you need it. To see how powerful the use of macros can be, perform the following steps:

1. Clear the contents of all of the cells in columns D and E. First select columns D and E by clicking on the column headings, and then right-click and select **Clear Contents** from the Quick Edit menu.
2. Select the top value in the data set, cell B4.
3. Run your macro with the following Ribbon options: **View tab → Macros group → Macros drop-down menu → Select the View Macros option.** [Excel 2003: Tools → Macros → Macros.] The Macro dialog box will be displayed (Figure 3.65).

**Figure 3.65**
The Macro dialog box.

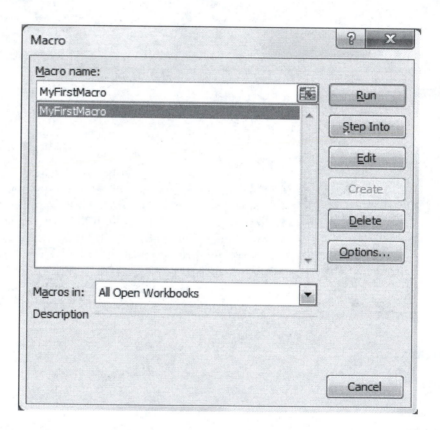

4. Select your macro's name from the list and then click the **Run** button. Voilà! Your worksheet will automatically re-perform all of the commands that you previously recorded.

### 3.8.3 Editing a Macro

Several functions may be performed from the Macro dialog box (Figure 3.65).

- The macro may be deleted by choosing the **Delete** button.
- A shortcut key may still be added for a macro by choosing the **Options** button.
- The **Step Into** button allows you to debug a macro.
- You can edit the macro (stored in a Visual Basic module).

Select your macro name from the list and choose the **Edit** button on the Macro dialog box. The Microsoft Visual Basic editor (where you can create, modify, and manage your macros) will open as depicted in Figure 3.66. Inside the editor, you can view and modify the Visual Basic code that was created when you recorded the macro.

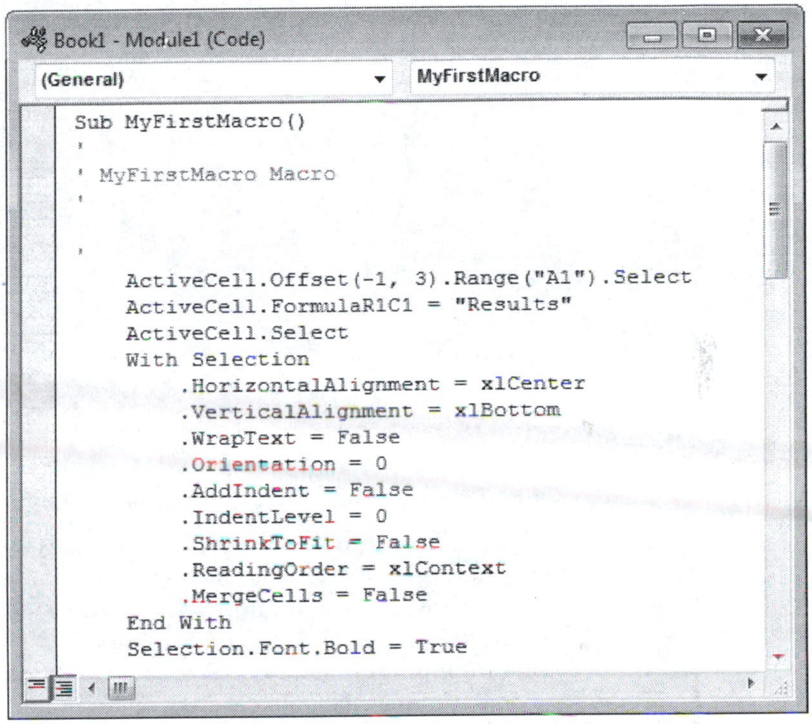
Book1 - Module1 (Code)
(General)
MyFirstMacro
Sub MyFirstMacro()
'
' MyFirstMacro Macro
'
'
ActiveCell.Offset(-1, 3).Range("A1").Select
ActiveCell.FormulaR1C1 = "Results"
ActiveCell.Select
With Selection
.HorizontalAlignment = xlCenter
.VerticalAlignment = xlBottom
.WrapText = False
.Orientation = 0
.AddIndent = False
.IndentLevel = 0
.ShrinkToFit = False
.ReadingOrder = xlContext
.MergeCells = False
End With
Selection.Font.Bold = True

**Figure 3.66**
The macro code in the Visual Basic module.

For example, the following two lines of code placed the text "Results" in cell E3:

ActiveCell.Offset(−1, 3).Range("A1").Select

ActiveCell.FormulaR1C1 = "Results"

## PRACTICE!

Convince yourself that the code displayed in the Visual Basic editor is actually the same code contained in the macro. You can do this without knowing the Visual Basic language. Make a couple of trivial changes, such as changing the line

ActiveCell.FormulaR1C1 = "Mean:"

to

ActiveCell.FormulaR1C1 = "Average:"

Execute the modified macro by choosing the Run button on the Visual Basic editor standard toolbar. Your worksheet should change to reflect your code change as illustrated in Figure 3.67!

| | A | B | C | D | E | F |
|---|---|---|---|---|---|---|
| 1 | Statistical Calculations | | | | | |
| 2 | | | | | | |
| 3 | | Data | | | Results | |
| 4 | | 10 | | Average: | 29.5 | |
| 5 | | 12 | | Median: | 23 | |
| 6 | | 45 | | Std. Dev.: | 19.24 | |
| 7 | | 32 | | | | |
| 8 | | 23 | | | | |
| 9 | | 23 | | | | |
| 10 | | 76 | | | | |
| 11 | | 21 | | | | |
| 12 | | 32 | | | | |
| 13 | | 21 | | | | |
| 14 | | | | | | |

**Figure 3.67**
The macro results after modifying the Visual Basic code.

## CREATING TABLES OF COMPOUND AMOUNT FACTORS

Excel is a great tool for creating and displaying tables of information based upon repetition of formulas. The next example shows how to create a table that shows the power of compound interest—a *compound amount factor* table. By using this table, an investor can see how the value of an initial investment will increase over time at different interest rates. For example, a person who invests $1,000 at 6% interest for 20 years will have $1,000 × 3.2017 or $3,207. A person who invests $1,000 at 11% interest for 20 years will have $1,000 × 8.0623, or $8,062.

Because investments with higher interest payments normally carry a higher degree of risk, such a table helps us assess whether the potential increase in payoff makes the risk acceptable. To create the table, follow these steps:

1. Enter titles and headings as shown in Figure 3.68.

| | A | B | C | D | E | F | G | H | I |
|---|---|---|---|---|---|---|---|---|---|
| 1 | Compound Amount Factors | | | | | | | | |
| 2 | | | | | | | | | |
| 3 | Interest Rate: | | 6% | 7% | 8% | 9% | 10% | 11% | |
| 4 | | | | | | | | | |
| 5 | | Years | | | | | | | |
| 6 | | | | | | | | | |

**Figure 3.68**
Titles and headings.

2. Create the column of values indicating the duration of the investment in years. First, enter 0 in cell B6 and 10 in cell B7. Then select the region B6:B7 and drag the Fill Handle down seven more rows. For appearance, center justify the year values as shown in Figure 3.69.

| | A | B | C | D | E | F | G | H | I |
|---|---|---|---|---|---|---|---|---|---|
| 1 | Compound Amount Factors | | | | | | | | |
| 2 | | | | | | | | | |
| 3 | Interest Rate: | | 6% | 7% | 8% | 9% | 10% | 11% | |
| 4 | | | | | | | | | |
| 5 | | Years | | | | | | | |
| 6 | | 0 | 1.0000 | 1.0000 | 1.0000 | 1.0000 | 1.0000 | 1.0000 | |
| 7 | | 10 | | | | | | | |
| 8 | | 20 | | | | | | | |
| 9 | | 30 | | | | | | | |
| 10 | | 40 | | | | | | | |
| 11 | | 50 | | | | | | | |
| 12 | | 60 | | | | | | | |
| 13 | | | | | | | | | |

**Figure 3.69**

Investment durations added in the Years column, and baseline factors (1.0000 in year zero) entered.

3. Type in 1 for year zero under each interest rate (cells C6:H6). Format region C6:H12 to display four decimal places. The result is shown in Figure 3.69.
4. The formula for single-payment compound factor is $[1 + i]^N$, where $i$ is the interest rate and $N$ is the number of interest periods—in this case, years. Enter the following formula for the single-payment compound amount factor in cell C7:

$$= (1 + C\$3)\char`\^\$B7$$

Notice that dollar signs have been included to indicate that the interest rate is always in row 3 and the duration (number of interest periods) is always in column B.

5. Use the Fill Handle to copy the formula to cells C8:C12.
6. Select the region C7:C12. Drag the Fill Handle to the right until the whole table is populated. Your completed table should resemble Figure 3.70.

| | A | B | C | D | E | F | G | H | I |
|---|---|---|---|---|---|---|---|---|---|
| 1 | Compound Amount Factors | | | | | | | | |
| 2 | | | | | | | | | |
| 3 | Interest Rate: | | 6% | 7% | 8% | 9% | 10% | 11% | |
| 4 | | | | | | | | | |
| 5 | | Years | | | | | | | |
| 6 | | 0 | 1.0000 | 1.0000 | 1.0000 | 1.0000 | 1.0000 | 1.0000 | |
| 7 | | 10 | 1.7908 | 1.9672 | 2.1589 | 2.3674 | 2.5937 | 2.8394 | |
| 8 | | 20 | 3.2071 | 3.8697 | 4.6610 | 5.6044 | 6.7275 | 8.0623 | |
| 9 | | 30 | 5.7435 | 7.6123 | 10.0627 | 13.2677 | 17.4494 | 22.8923 | |
| 10 | | 40 | 10.2857 | 14.9745 | 21.7245 | 31.4094 | 45.2593 | 65.0009 | |
| 11 | | 50 | 18.4202 | 29.4570 | 46.9016 | 74.3575 | 117.3909 | 184.5648 | |
| 12 | | 60 | 32.9877 | 57.9464 | 101.2571 | 176.0313 | 304.4816 | 524.0572 | |
| 13 | | | | | | | | | |

**Figure 3.70**

The completed Single-Factor, Compound Amount Factor table.

Notice that the equation for a compound amount factor was entered only once, in cell C7. All that was necessary to complete the rest of the table was to copy the equation in cell C7 to the rest of the cells in the table.

## INTERACTIVE DC CIRCUIT ANALYZER

When there is a sudden change in a DC circuit that includes a capacitor or an impedance, the current in the circuit varies with time (e.g., until the capacitor is charged). A general expression for the current $I$ in a DC transient circuit is

$$I(t) = I_\infty + (I_0 - I_\infty)e^{-t/T},$$

where

$I_0$ is an initial value at the instant of sudden change,
$I_\infty$ is the current at time $t = \infty$,
$T = RC$ is the time constant for a series $R$–$C$ (resistance-capacitance) circuit,

and

$T = L/R$ is the time constant for a series $R$–$L$ (resistance-impedance) circuit.

**Figure 3.71**
An $R$–$L$ circuit.

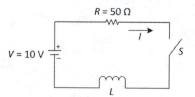

If switch $S$ is closed at $t = 0$, then we can calculate the current in the circuit as a function of time. In this example we will determine the current in the circuit after a time equal to three time constants ($t = 3T$). $I_0 = 0$ in this example (no current when the switch was open). We will create a worksheet that calculates $I(t = 3T)$ for any values of $V$ and $R$ entered by a user. If $I_0$, $V$, and $R$ are entered into cells B4, B5, and B6, respectively, then the current at $t = \infty$ can be found from $I_\infty = V/R$. In our worksheet, the Excel formula for $I_\infty$ is

$$= B5/B6,$$

and the Excel formula for $I(t = 3T)$ is

$$= B9+(B4-B9)*EXP(B10)$$

If you enter the voltage (10 volts) and resistance (50 ohms) from the circuit displayed in Figure 3.71, then your calculated results should resemble those in Figure 3.72. The current at three time constants after the switch is closed equals 0.19 amps.

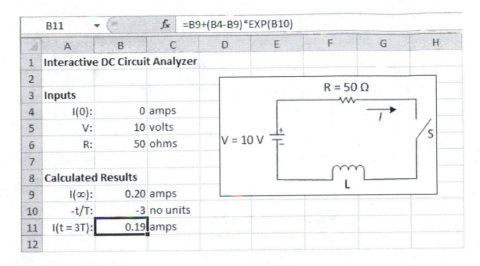

**Figure 3.72**
Interactive calculator for finding the current in a DC transient circuit.

You can use a worksheet like the one shown in Figure 3.72 to compute the current interactively by entering various values for $V$ and $R$.

## KEY TERMS

absolute cell referencing
*ACOS* function
*ASIN* function
*ATAN* function
*AVERAGE* function
built-in functions
cell range
compound amount factor
Ctrl + Shift + Enter
   (array functions)
debugging
degrees
*DEGREES* function
dependent tracing

error messages
Fill Handle
formula
Formula bar
future value
Insert Function button
macro
macro virus
matrix
matrix order
matrix product
*MEDIAN* function
*MMULT* function
Name box

named cell
named cell range
operator precedence
precedence
precedent tracing
present value
prototype
radians
relative cell referencing
Ribbon
*STDEV* function
*SUM* function
syntax
transpose (matrix)

## SUMMARY

### Referencing Cells

- Relative reference—cell addresses follow the formula when copied.
- Absolute reference—cell addresses are fixed (use $ to create an absolute reference).

- Examples
  - $A$1    both row and column are absolute
  - $A1      column is absolute, row is relative
  - A$1      column is relative, row is absolute

### Naming Cells and Cell Ranges

**1.** Select the cell or cell range.

**2.** Enter the name in the Name Box (left side of Formula bar).

Use the Name Manager [**Formula tab → Defined Names group → Name Manager**] to remove a name from a cell or cell range.

### Formula Syntax

- Start a formula with an equal sign (=).
- Formulas can contain:
  - Math operators
  - Values
  - Cell references
  - Function calls

### Operator Precedence

- Formulas are generally interpreted from left to right, except:
  - Operator precedence rules (see the following table) are enforced.
  - Parentheses can be used to control how a function is interpreted.

| Precedence | Operator | Operation |
|:---:|:---:|:---|
| 1 | % | Percentage |
| 2 | ^ | Exponentiation |
| 3 | *, / | Multiplication, Division |
| 4 | +, − | Addition, Subtraction |

### Built-In Functions

- Use the Insert Function button on the Formula bar to open the Insert Function dialog.
- Search for a function (from the Insert Function dialog):
  - Using a search
  - By category
- Function Arguments dialog boxes:
  - Identify all required and optional arguments.
  - Assist you in correctly using the functions.
- The Excel Help system provides information on built-in functions.

### Commonly Used Excel Functions (only a small portion of Excel's built-in functions are listed here)

### Basic Math Functions

- SUM
- LOG

- LN
- EXP
- FACT

### Statistical Functions

- AVERAGE
- STDEV
- MEDIAN

### Trigonometric Functions (angles are measured in radians)

- SIN
- COS
- TAN
- ASIN
- ACOS
- ATAN
- DEGREES (converts radians to degrees)
- RADIANS (converts degrees to radians)

### Matrix Functions (enter an array function with **Ctrl + Shift + Enter**)

- MMULT
- MINVERSE
- MDETERM
- TRANSPOSE

### Financial Functions

- PV
- FV
- PMT

### Excel Error Messages

| Message | Description |
| --- | --- |
| ###### | The value is too wide to fit in the cell, or an attempt was made to display a negative date or time. |
| #VALUE | The wrong type of argument was used in a formula. This will occur, for example, if text were entered when an array argument was expected. |
| #DIV/0 | An attempt was made to divide by zero in a formula. See the quadratic equation example in Figure 3.50 for an illustration. |
| #NAME | A name used in a formula is not recognized. Usually, the function or defined name was misspelled. Note that named ranges or functions may not contain spaces. |
| #REF | A referenced cell is not valid. This usually occurs when a cell is referenced in a formula and that cell is then deleted. It also occurs if an attempt is made to paste a cell over a referenced cell. |
| #NUM | The expression produces a numeric value that is out of range or invalid. Examples are extremely small, large, or imaginary numbers. To see this error, try this formula: =SQRT(-1). |
| #NULL | An attempt was made to reference the intersection of two areas that don't intersect. This usually occurs when a space is inadvertently placed between two arguments, instead of a comma or colon, as in =SUM(C2 D3). |

**Debugging Worksheets**

- Highlight Formulas [**Home tab → Editing group → Find & Select drop-down menu → Formulas option**]
- Display Formulas [**Formulas tab → Formula Auditing group → Show Formulas button**]
- Tracing Dependents [**Formulas tab → Formula Auditing group → Trace Dependents button**]
- Tracing Precedents [**Formulas tab → Formula Auditing group → Trace Precedents button**]
- Remove Arrows [**Formulas tab → Formula Auditing group → Remove Arrows button**]
- Pattern Errors—Excel marks the corners of cells that appear not to match the pattern of adjacent cells

**Macros**

- Recording a Macro [**View tab → Macros group → Macros drop-down menu → Record Macro . . . button**]
- Running a Macro [**View tab → Macros group → Macros drop-down menu → Select the View Macros option**, then select the desired macro and click the **Run** button]
- Activating Relative Referencing [**View tab → Macros group → Macros drop-down menu → Click the Use Relative References toggle button**]

## PROBLEMS

1. Place the number 10 in cell A1. Create an Excel formula that computes

$$f(x) = x^2 - 4x + 3$$

using the value in cell A1 as $x$.

2. Place the values for $x = 5$ and $y = 7$ in cells A1 and A2, respectively. Create an Excel formula that computes

$$f(x, y) = y^3 - 10x^2$$

using the values in cells A1 and A2.

3. Place the numbers 1, 2, . . . 10 in cells A1:A10. Create an Excel formula in cell B1 that computes

$$f(x) = \ln x + \sin x$$

using the value in cell A1. Use the Fill Handle and drag the formula over cells B2:B10 to evaluate $f(x)$ for all 10 values.

4. For a damped oscillation as depicted in Figure 3.73, the displacement of a structure is defined by the equation

$$f(t) = 8e^{-kt} \cos(\omega t),$$

where $k = 0.5$ and the frequency, $w = 3$. Create an Excel formula for this equation. Compute $f(t)$ for $t = 0.0, 0.1, 0.2, . . . 4.0$ s. What is the value of $f(t)$ for $t = 3.6$ s?

**Figure 3.73**

Damped oscillation representing displacement of a structure.

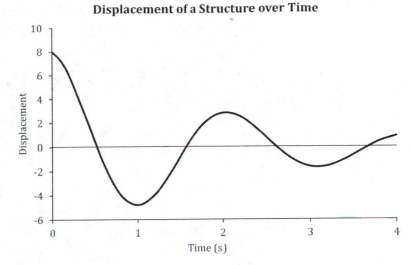

**Displacement of a Structure over Time**

5. The formula that calculates the number of combinations of $r$ objects taken from a collection of $n$ objects is

$$C(n, r) = \frac{n!}{(n - r)!\,r!}$$

The exclamation point is the mathematical symbol for the *factorial* operation. The factorial of a number $n = n \times (n-1) \times (n-2) \times , \ldots \times 3 \times 2 \times 1.$ Thus, the factorial of

$$4 = 4 \times 3 \times 2 \times 1 = 24$$

The Excel function for factorial is called *FACT*. The preceding formula can be used to compute the number of ways a committee of six people can be chosen from a group of eight people, as illustrated in Figure 3.74.

**Figure 3.74**

Calculating the number of possible combinations.

| C6 | | | $f_x$ | =FACT(C4)/(FACT(C4-C3)*FACT(C3)) | | |
|---|---|---|---|---|---|---|
| | A | B | C | D | E | F |
| 1 | Combinations | | | | | |
| 2 | | | | | | |
| 3 | | r: | 6 | | | |
| 4 | | n: | 8 | | | |
| 5 | | | | | | |
| 6 | | combinations: | 28 | | | |
| 7 | | | | | | |

Write an Excel workbook to calculate combinations. Use it to compute how many 5-card hands may be drawn from a deck of 52 cards.

6. Excel has a number of predefined logical functions. One of these, the *IF* function, has the following syntax:

$$= IF(TEST, T, F)$$

The effect of the function is to evaluate the expression, *TEST*, which must be a logical expression. If the expression evaluates to "true," then $T$ is returned. If the expression valuates to "false," then $F$ is returned. For example, the *IF* function

$$IF (X < 200, X, \text{"Cholesterol is too high"})$$

tests to see if $X$ is less than 200 (see Figure 3.75).

- If $X$ is less than 200, the *IF* function returns the value of $X$.
- If $X$ is greater than, or equal to, 200, then the text statement "Cholesterol is too high" is returned (and displayed in the cell containing the *IF* function).

**Figure 3.75**
Using the *IF* function.

| | A | B | C | D | E | F |
|---|---|---|---|---|---|---|
| | | | | | fx | =IF(C3<200,C3,"Cholesterol is too high") |
| 1 | Cholesterol Check | | | | | |
| 2 | | | | | | |
| 3 | | X: | 210 | | | |
| 4 | | Check: | Cholesterol is too high | | | |
| 5 | | | | | | |

C4 cell reference shown above.

To try the IF function, expand the quadratic equation example (see Figure 3.50) to test for division by zero. If the expression $2a = 0$ is true, then display "Divide by Zero"; otherwise, return the value of $2a$.

Perform a similar test for $b^2 - 4ac \geq 0$. Display "Requires Complex Number" if the test is false.

7. Neglecting air resistance, the horizontal range of a projectile fired into the air at angle $\theta$ degrees is given by the formula

$$R = \frac{2V^2 \sin \theta \cos \theta}{g}$$

Create a worksheet that computes $R$ for a selected initial velocity $V$ and firing angle $\theta$. Use $g = 9.81$ m/s². Convert degrees to radians by using the *RADIANS* function. To test your results, an initial velocity of 150 m/s and firing angle of 25° should result in $R = 1,756$ m.

8. Two frequently performed matrix operations are the calculation of the *determinant* of a matrix and the *inverse* of a matrix. The Excel functions for these operations are *MDETERM* and *MINVERSE*, respectively. Create matrices A and B as shown in Figure 3.32. Compute the matrix inverse and determinant of A and B.

9. Create a macro that computes, labels, and displays the determinant and inverse of a $3 \times 3$ matrix which is typed into cells A1:C3. Create a shortcut key to execute the macro.

10. Using trigonometry, you can calculate the height of a tall tree on a sunny day by measuring the length of the tree's shadow and the angle that the line between the tip of the shadow and the tip of the tree makes with the ground. The required variables are indicated in Figure 3.76.

If the length of the shadow is 22 m and the angle is 49°, the height can be determined by using the tangent function, as

$$height = L\, TAN(49°).$$

But remember, Excel's *TAN* function wants the angle in radians, not degrees. Use the *RADIANS* function to convert 49° to radians, then use Excel's *TAN* function to determine the height of the tree.

**Figure 3.76**
Determining the height of
a tree on a sunny day.

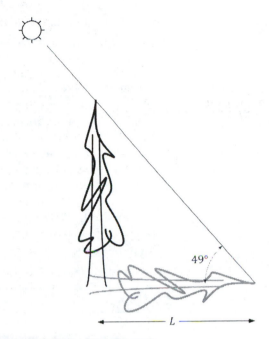

11. If a groundskeeper wants to apply 0.022 kg of fertilizer to each square meter of a football field, how much fertilizer should he or she purchase? Create an Excel worksheet to determine the answer.

**Data**

- A football field is 100 yards long and 50 yards wide.
- There are 3 feet per yard.
- There are 3.2808 feet per meter.

12. Any time you want to borrow money from a bank, credit union, or car dealer, you will have to pay back the loan in monthly payments with interest. The interest rate will be stated as an *annual percentage rate*, or *APR*. However, you will need a monthly interest rate to calculate the monthly payment. The monthly interest rate is simply the APR/12.

Create your own loan payment calculator in Excel. An example for a $10,000 loan with a 7% APR is shown in Figure 3.77. For this loan, the monthly payment will be about $309.

**Figure 3.77**
Loan payment calculator.

| C8 | | fx | =PMT(C5,C6,C3) | | |
|---|---|---|---|---|---|
| ▲ | A | | B | C | D |
| 1 | Loan Payment Calculator | | | | |
| 2 | | | | | |
| 3 | | | Amount borrowed (P): | $10,000 | |
| 4 | | | Annual percentage rate (APR): | 7% | |
| 5 | | | Monthly interest rate (i): | 0.58% | |
| 6 | | | Number of monthly payments: | 36 | |
| 7 | | | | | |
| 8 | | | Amount of each payment (PMT): | -$308.77 | |
| 9 | | | | | |

Follow these steps to create the loan payment calculator worksheet:

1. Enter the labels as shown in column B in Figure 3.77.
2. Enter an amount in cell C3. This is the amount you want to borrow. Format cell C3, using a Currency or Accounting format.
3. Enter the APR (from the bank) in cell C4. Format this cell with a Percentage format. (If you type 7% into the cell, Excel will automatically apply a Percentage format.)
4. In cell C5, enter the formula =C4/12 to compute the monthly interest rate from the annual rate. Format cell C5 with a Percentage format and display at least two decimal places.
5. Enter the number of monthly payments in cell C6.
6. Use Excel's *PMT* function in cell C8 to compute the required annual payment. The *PMT* function will use the values in cells C3, C5, and C6 as arguments. Use the Insert Function dialog box or the Help system, to determine the proper syntax of the *PMT* function arguments.

Calculate the required payment for each of the following types of loans:

| Loan Type | P | APR (%) | N (months) |
|---|---|---|---|
| Car loan | $ 21,000 | 8.0 | 48 |
| Student loan | $ 24,000 | 4.3 | 120 |
| Home loan | $250,000 | 6.5 | 360 |

13. Calculate the mean, median, and mode for the grade data shown in Figure 3.78. The Excel functions used to compute these statistics are the *AVERAGE, MEDIAN,* and *MODE* functions, respectively.

**Figure 3.78**
Course scores.

| | A | B | C |
|---|---|---|---|
| 1 | Grade Data | | |
| 2 | | | |
| 3 | Student | Score | |
| 4 | 1 | 78 | |
| 5 | 2 | 88 | |
| 6 | 3 | 98 | |
| 7 | 4 | 88 | |
| 8 | 5 | 100 | |
| 9 | 6 | 95 | |
| 10 | 7 | 82 | |
| 11 | 8 | 96 | |
| 12 | 9 | 82 | |
| 13 | 10 | 98 | |
| 14 | 11 | 84 | |
| 15 | 12 | 64 | |
| 16 | 13 | 82 | |
| 17 | 14 | 77 | |
| 18 | 15 | 92 | |
| 19 | | | |
| 20 | mean: | | |
| 21 | median: | | |
| 22 | mode: | | |
| 23 | | | |

14. A discount store sells two different brands of tire pressure gauges. Joseph wants to get a tire gauge, but wants to get the best one he can. He convinces the store manager to let him try two brands to see which one works better. He goes out to the parking lot and takes 10 pressure readings with each gauge on a single tire. The results (pressures in $lb_f/in^2$ or psi) are shown in Figure 3.79.

**Figure 3.79**

Pressure readings on a single tire.

| | A | B | C |
|---|---|---|---|
| 1 | Comparing Pressure Gauges | | |
| 2 | | | |
| 3 | Gauge A | Gauge B | |
| 4 | 28 | 23 | |
| 5 | 28 | 32 | |
| 6 | 28 | 23 | |
| 7 | 25 | 27 | |
| 8 | 26 | 25 | |
| 9 | 26 | 32 | |
| 10 | 26 | 29 | |
| 11 | 28 | 23 | |
| 12 | 29 | 26 | |
| 13 | 28 | 29 | |
| 14 | | | |

Use Excel's *AVERAGE* and *STDEV* functions to calculate the average pressure and the standard deviation for each column of data. Which gauge do you think Joseph should buy?

15. Matrix math can be used to solve systems of linear equations such as the three equations shown as follows:

$$1x_1 + 3x_2 + 5x_3 = 6$$
$$1x_1 + 4x_2 + 7x_3 = 4$$
$$2x_1 + 3x_2 + 1x_3 = 9$$

To do so, the coefficients multiplying the $x$ variables are collected into a coefficient matrix, $C$, and the constants on the right side of the equations are collected in a vector (single column matrix), $r$.

$$C = \begin{bmatrix} 1 & 3 & 5 \\ 1 & 4 & 7 \\ 2 & 3 & 1 \end{bmatrix} \qquad r = \begin{bmatrix} 6 \\ 4 \\ 9 \end{bmatrix}$$

The $x$ values ($x_1$, $x_2$, and $x_3$) are determined by inverting the $C$ matrix and multiplying that result by the $r$ vector:

$$x = C^{-1}r$$

In Excel, the process is carried out by using the *MINVERSE* function and the *MMULT* function. The results are shown in Figure 3.80.

**Figure 3.80**

Solving simultaneous linear equations.

| | A | B | C | D | E | F | G | H |
|---|---|---|---|---|---|---|---|---|
| 1 | Solving Simultaneous Linear Equations | | | | | | | |
| 2 | | | | | | | | |
| 3 | [C] | 1 | 3 | 5 | | [r] | 6 | |
| 4 | | 1 | 4 | 7 | | | 4 | |
| 5 | | 2 | 3 | 1 | | | 9 | |
| 6 | | | | | | | | |
| 7 | [C]inv | 5.666667 | -4 | -0.33333 | | [x] | 15 | |
| 8 | | -4.33333 | 3 | 0.666667 | | | -8 | |
| 9 | | 1.666667 | -1 | -0.33333 | | | 3 | |
| 10 | | | | | | | | |

**a.** Verify that the calculated results $x_1 = 15$, $x_2 = -8$, and $x_3 = 3$ do satisfy the simultaneous equations.

**b.** Modify the worksheet to solve the following set of simultaneous equations:

$$1x_1 + 3x_2 + 5x_3 = 3$$
$$1x_1 + 4x_2 + 7x_3 = 5$$
$$2x_1 + 3x_2 + 1x_3 = 12$$

# 4

# Working with Charts

## Sections

## Objectives

*After reading this chapter, you should be able to perform the following tasks:*

- Insert a basic chart on a worksheet.
- Use a chart layout to add additional features.
- Create Line charts and XY Scatter charts.

- Format chart legends, axes, and titles.
- Preview and print a chart.
- Add an additional data series to a chart.
- Scale axes and create error bars.

## 4.1 INTRODUCTION

Excel 2007 introduced a new procedure for creating a basic graph, or *chart*, from a data set. The new procedure is still used with Excel 2010; you create a basic chart from your data set, and then refine the chart's features using various Ribbon options. When you are working with a chart, three new Ribbon tabs are displayed to provide access to chart formatting options. These will be presented in detail later in the chapter.

The organization of the first sections of this chapter follows the typical steps that you will use in preparing a chart:

- Create the basic chart (Section 4.3).
- Select a chart layout (Section 4.3).
- Format features of the basic chart (Section 4.4).
- Preview and print the chart (Section 4.6).

Then, additional information is presented on ways to add additional data to an existing chart and how to prepare a chart with more than one curve. Finally, there is a discussion of certain chart features that are particularly useful to engineers.

## 4.2 CREATING A BASIC XY SCATTER CHART

Creating a basic chart in Excel is trivial:

1. Select the data to be charted.
2. Use the Ribbon's Insert tab.
3. Select the type of chart to be created from the Charts group.

That's it—the basic chart will be created and placed on the worksheet. Once the basic chart is created you can start modifying it to meet your needs.

Creating a chart in Excel is nearly automatic, but you need to understand that Excel is making the process easy by making some assumptions about how your data is organized. Your charting will be less stressful if you organize the data to be charted in ways that Excel can recognize.

- The data values for each *series* (each curve on a chart) should be in a single column or row. Columns are more common, but rows will work.
- If you want to name a data series, the name should be in a single cell at the top of the column or left end of the row.
- For an XY Scatter chart, which requires both $x$ and $y$ values for plotting, the $x$ values should be in the left column, or top row.

Figure 4.1 shows a typical data layout for creating an XY Scatter chart. The data to be plotted are in cells A3:B12, including the column headings. The potential values will be plotted on the **x-axis**, and the current values on the **y-axis**.

**Figure 4.1**
Data collected by measuring current across a 150 Ω resistor.

| | A | B | C |
|---|---|---|---|
| 1 | Current - Potential Data | | |
| 2 | | | |
| 3 | Potential (V) | Current (A) | |
| 4 | 6.97 | 0.051 | |
| 5 | 5.96 | 0.044 | |
| 6 | 4.95 | 0.038 | |
| 7 | 3.98 | 0.032 | |
| 8 | 3.03 | 0.025 | |
| 9 | 1.91 | 0.018 | |
| 10 | 1.02 | 0.012 | |
| 11 | 0.50 | 0.008 | |
| 12 | 0.20 | 0.001 | |
| 13 | | | |

In this chapter we will cover XY Scatter charts first, because they are very commonly used in engineering.

### 4.2.1 Creating a Basic XY Scatter Chart

An *XY Scatter chart* shows the relationship between two variables, one plotted on the *x*-axis and the other on the *y*-axis. They are useful for visualizing relationships among data; how one variable responds to changes in the other variable.

Before proceeding, create the worksheet shown in Figure 4.1 so that you can work through the examples in this chapter. The data in Figure 4.1 were collected by measuring current (*I*) in amperes (*A*) across a resistor for nine measured voltages (*V*). Ohm's law, which is

$$V = IR,$$

states that the relationship between *V* and *I* is linear if temperature is kept relatively constant. In this section we will create a basic XY Scatter chart from the data in Figure 4.1. An XY Scatter chart locates points using the *x* and *y* values associated with each data point. Because the *x* and *y* values for each point on the chart are paired, they are sometimes referred to as *x, y data pairs*.

Use the following steps to create an XY Scatter chart of the data in Figure 4.1:

1. Select the data to be charted (including the headings, if desired), cells A3:B12 in this example.
2. Insert the basic XY Scatter chart on your worksheet using Ribbon options: **Insert tab → Charts group → Scatter drop-down menu → Scatter with only markers style**. These Ribbon options are illustrated in Figure 4.2.

**Figure 4.2**

Inserting a basic XY Scatter chart on the worksheet.

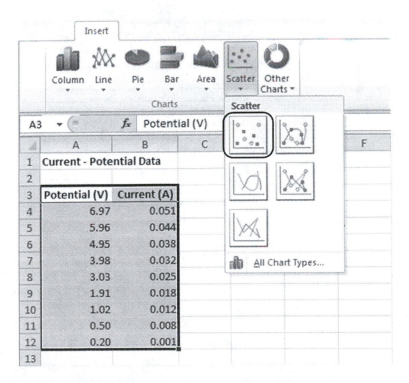

Notice, in Figure 4.2, that there are five styles of XY Scatter charts available on the Scatter drop-down menu:

- *Markers* only, no connecting *lines*
- Markers with smoothed connecting lines
- *Smoothed connecting lines*, no markers
- Markers and connecting lines, no line smoothing
- Connecting lines, no line smoothing, no markers

When inserting the basic chart, simply select the style you want Excel to use to create the chart. In this example we will use markers only. The basic XY Scatter chart created from our data is shown in Figure 4.3.

**Figure 4.3**

The basic XY Scatter chart on the worksheet.

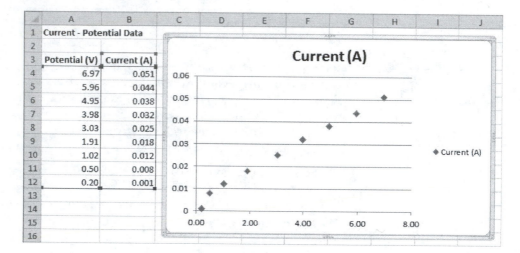

Creating the basic chart is very easy, but there are no *axis labels* yet, and you may want to change the default title. Those changes will be described in the next section.

Notice some worksheet features (see Figure 4.3) that can be useful when working with charts:

- When a chart is selected, as in Figure 4.3, it is shown with a heavy border. If you need to move the chart, grab the border with the mouse and drag the chart to a new location.
- The chart border includes *handles* (three small dots) on each corner and side. Those handles can be used (with the mouse) to change the size of the displayed chart.
- The data used to create the chart is indicated by colored boxes (around cells A4:A12 and B4:B12).
- The series heading that was selected before creating the graph ("Current (A)" in cell B3) was used as the series title in the legend (right side of chart) and as the chart title (top of chart).

The next step in creating a useful chart is getting axis labels on the chart. This is accomplished by selecting a chart layout. This is the subject of the next section.

## 4.3 SELECTING A CHART LAYOUT

Whenever a chart is selected, Excel makes additional chart tools available on the Ribbon in three new Chart Tool tabs:

- *Design* – global aspects of the chart's design (chart type, layout, basic style).
- *Layout* – individual elements of a chart (axis labels, *gridlines*).
- *Format* – specific format details (shape line color, font color).

The Design tab (Figure 4.4) is the most general, while the Format tab gets much more detailed. You typically start editing your chart with the Design tab.

**Figure 4.4**
The Chart Tools: Design tab on the Ribbon when a chart is selected.

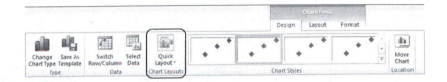

When a basic XY Scatter chart is created it is missing some key features, primarily features used to describe what variable is being plotted on each axis, and the units on the variable. The quickest way to add features to a chart is to select a *chart layout*. A variety of chart layouts are available from the Ribbon: **Chart Tools → Design tab → Chart Layouts group → Chart Layouts palette**. The Chart Layouts palette is shown in Figure 4.5. This is a very complex selection palette allowing you to choose various options, including:

- Axis labels
- Legend location
- Gridline style
- Linear trendline

The good news is that Layout 1 (indicated in Figure 4.5) works for a lot of XY Scatter chart applications.

**Figure 4.5**
Chart Layouts palette.

Layout 1 ⟶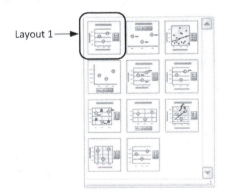

To apply a chart layout:

1. Select the chart by clicking on some white space near the edge of the chart. You want the entire chart selected, not any of the elements of the chart such as an axis, a title, or a data point.
2. Click on the desired chart layout: **Chart Tools → Design tab → Chart Layouts group → Chart Layouts palette → select the desired layout.**

In Figure 4.6 Layout 1 has been applied to the chart. The only real change is that the words "Axis Title" now appear on the *x* and *y* axes.

**Figure 4.6**
The XY Scatter chart after applying Layout 1.

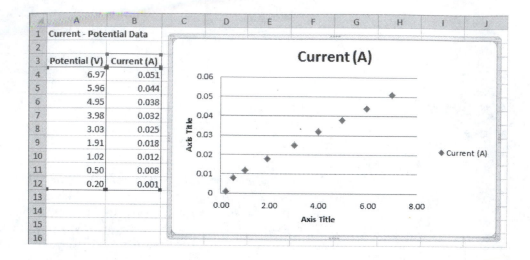

To change the text of either axis title, or the chart title, simply select the title by clicking on it with the mouse, and start typing. The new text appears in the Formula bar (see Figure 4.7) until you press **Enter** and then the chart is updated.

**Figure 4.7**
Entering an axis title using the Formula bar for text entry.

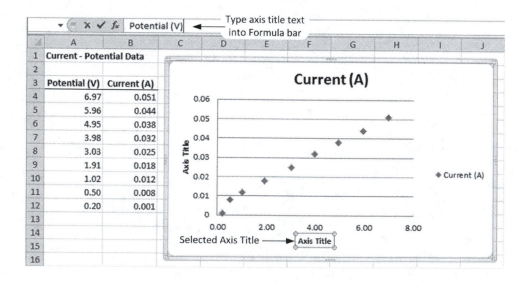

Alternatively, you can edit a title "in place" with two careful mouse clicks (not a double-click) on the title (Figure 4.8):

1. Click once to select the title.
2. Click again to edit in place.

**Figure 4.8**
Editing a title in place.

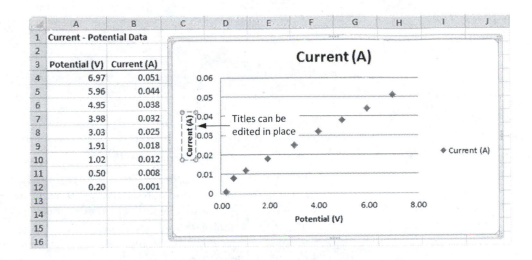

If there is a feature that you don't want, click on the feature to select it and then press **Delete**. For example, a *legend* is used to help the reader understand which curve is associated with which data series. A legend isn't really necessary when there is a single curve on the chart (as is the case in this example), so it can be deleted, as shown in Figure 4.9.

**Figure 4.9**
The XY Scatter chart after updating titles and deleting the legend.

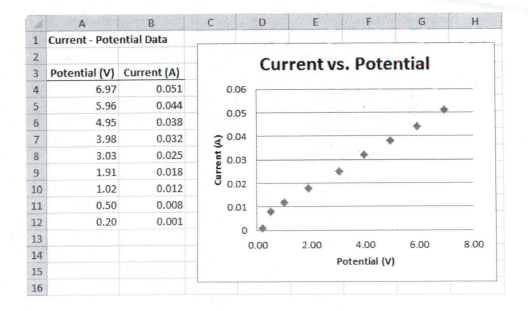

## 4.4 FORMATTING CHART FEATURES

After updating the chart layout, the chart is nearly ready to use. But the Layout tab on the Ribbon's Chart Tools (shown in Figure 4.10) allows you to update additional features, if needed.

**Figure 4.10**
The Chart Tools: Layout tab
on the Ribbon when a
chart is selected.

The Labels and Axes groups on the Layout tab allow you to quickly add chart features such as axis labels, a legend, or gridlines. If you want to annotate a chart, use a *text box* from the Insert group, as follows:

1. Select the chart (this causes the Chart Tools tabs to be available on the Ribbon).
2. Click the **Text Box** button: **Chart Tools → Layout tab → Insert group → Text Box button**. The mouse cursor will change to a vertical bar with a small crossbar showing the baseline of the text.
3. Position the mouse cursor on the chart where you want to place the text, and click the mouse. A text box will be placed on the chart and text entry will be initiated.
4. Type the desired text, as illustrated in Figure 4.11.
5. Click outside the text box to complete the text entry.

**Figure 4.11**
Adding text to a chart.

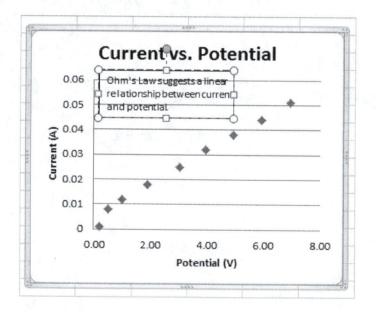

The text box is now an object associated with the chart, and can be further modified. For example, we can assign a white background to the text box and a dark border using the following steps:

1. Click the text box to select it. This causes the Drawing Tools: Format tab to be displayed on the Ribbon (Figure 4.12).
2. Use the Shape Fill drop-down menu (indicated in Figure 4.12) to select a white background for the text box.
3. Use the Shape Outline drop-down menu (indicated in Figure 4.12) to select a dark color for the text box's border.

The result is shown in Figure 4.13.

**Figure 4.12**
The Drawing Tools: Format tab is displayed when a text box has been selected.

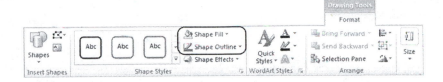

**Figure 4.13**
The annotated chart.

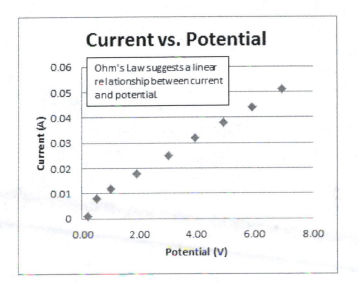

### 4.4.1 Editing an Existing Chart Feature

In order to edit an existing chart feature, you will usually need to open a dialog box for that feature. There are several ways to access these formatting dialog boxes:

- Use the drop-down menus on the Ribbon, and select the "More" option at the bottom of the menu.
- Select the feature to be modified and click the **Format Selection** button on the Ribbon's Layout (or Format) tab: **Chart Tools → Layout (or Format) tab → Current Selection group → Format Selection button**.
- Right-click the feature and select the "Format" option for that feature from the pop-up menu.

As an example, we will give the ***plot area*** (the rectangle behind the data points) a dark border.

1. Select the plot area by clicking on one of the edges of the plot (e.g., one of the axis lines). Be careful to select the plot area and not gridlines or axis numeric values.
2. Click the **Format Selection** button on the Ribbon: **Chart Tools → Layout tab → Current Selection group → Format Selection button**. The Format Plot Area dialog box will be displayed as shown in Figure 4.14.
3. Choose the Border Color panel.
4. Select the Solid Line option.
5. Choose a color from the Color drop-down palette.
6. Click the **Close** button when you are done making changes to the plot area.

The result is shown in Figure 4.15.

**Figure 4.14**
The Format Plot Area
dialog box, Border Color
panel.

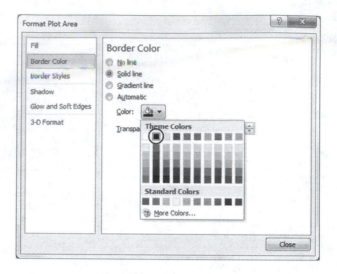

**Figure 4.15**
The reformatted chart.

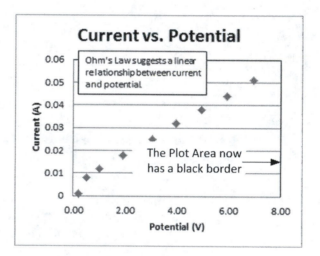

---

## PRACTICE!

By default, Excel 2010 uses the format of the plotted data values to format the numbers on the axis. This can result in more digits than needed displayed on an axis. For example, in Figure 4.15 the values on the *x*-axis are shown with two decimal places that don't need to be there. To get rid of the extraneous decimal places, we need to reformat the *x*-axis. Try it with the following steps:

1. Select the *x*-axis by clicking on any of the displayed axis numbers.
2. Click the **Format Selection** button on the Ribbon: **Chart Tools →Layout tab → Current Selection group → Format Selection button**. The Format Axis dialog box will be displayed as shown in Figure 4.16.
3. Choose the Number panel.
4. Set the **Decimal places** to zero as shown in Figure 4.16.
5. Click the **Close** button when you are done making changes to the axis.

The reformatted chart is shown in Figure 4.17.

**Figure 4.16**
The Format Axis dialog box.

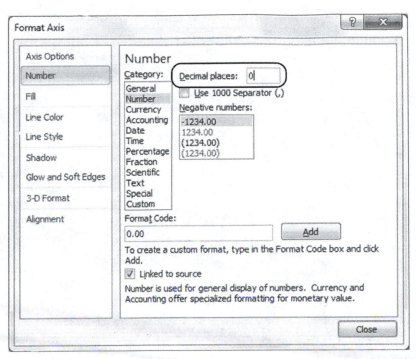

**Figure 4.17**
The chart after reformatting the x-axis.

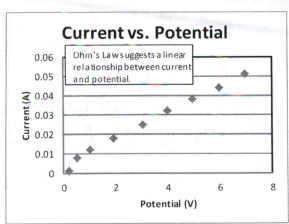

---

## PRACTICE!

Use the data given in Table 4.1 to practice creating an XY Scatter chart. Use a chart layout to get axis titles on the chart.

**Table 4.1  Tank Wash-Out Data**

| Time (s) | Conc. (mg/L) |
|----------|--------------|
| 0        | 100.0        |
| 10       | 36.8         |
| 20       | 13.5         |
| 30       | 5.0          |
| 40       | 1.8          |

Your chart should look something like Figure 4.18.

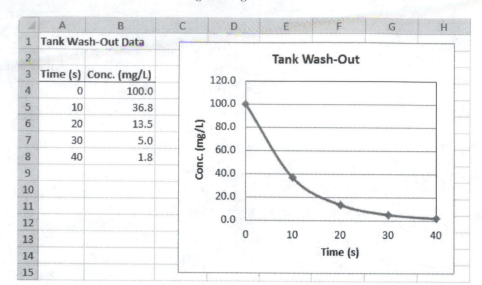

**Figure 4.18**
Plotting wash-out data.

## 4.5 CREATING A BASIC LINE CHART

The process for creating a Line chart is basically the same as for creating an XY Scatter chart:

1. Select the data to be charted (including the headings, if desired).
2. Insert the basic Line chart on your worksheet using the Ribbon options. **Insert tab → Charts group → Line drop-down menu → select a Line chart style**.

The difference between a **Line chart** and an XY Scatter chart in Excel is the way the $x$-axis is handled. In an XY Scatter chart, the $x$ location of each data point is along the $x$-axis according to the magnitude of the $x$ value associated with that point. In a Line chart there are no $x$ values, just $x$ labels. In a Line chart, the *labels* on the $x$-axis are simply distributed uniformly across the chart from left to right.

As an example of creating a Line chart, consider the data shown in Table 4.2. This data represents the low-flow rates for two tributaries of the Pecos River. The lowest one-day flow rate (in cubic feet per second, or cfs) for each year is shown. We will create a Line chart that shows how the flow rates vary from year to year. First, create a worksheet that contains the data in Table 4.2.

In this example, we are creating a Line chart with two data series, one for each tributary. A *data series* is a collection of related data points that are represented as a unit. Each data series will show up as a set of data points or a curve on the chart.

In preparation for creating a Line chart, the data in Table 4.2 is entered into an Excel worksheet, as shown in Figure 4.19.

**Table 4.2 Annual Low-Flow Rate of Pecos River**

| Year | East Branch Flow (cfs) | West Branch Flow (cfs) |
|---|---|---|
| 87 | 221 | 222 |
| 88 | 354 | 315 |
| 89 | 200 | 175 |
| 90 | 373 | 400 |
| 91 | 248 | 204 |
| 92 | 323 | 325 |
| 93 | 216 | 188 |
| 94 | 195 | 202 |
| 95 | 266 | 254 |
| 96 | 182 | 176 |

**Figure 4.19**
Entering the flow data into a worksheet.

|  | A | B | C | D | E |
|---|---|---|---|---|---|
| 1 | Annual Low-Flow Rate of the Pecos River | | | | |
| 2 | | | | | |
| 3 | Year | East Branch Flow (cfs) | West Branch Flow (cfs) | | |
| 4 | 87 | 221 | 222 | | |
| 5 | 88 | 354 | 315 | | |
| 6 | 89 | 200 | 175 | | |
| 7 | 90 | 373 | 400 | | |
| 8 | 91 | 248 | 204 | | |
| 9 | 92 | 323 | 325 | | |
| 10 | 93 | 216 | 188 | | |
| 11 | 94 | 195 | 202 | | |
| 12 | 95 | 266 | 254 | | |
| 13 | 96 | 182 | 176 | | |
| 14 | | | | | |

To create a basic Line chart with this data, follow these steps:

1. Select the data to be charted (including the headings), cells A3:C13 in this example.
2. Insert the basic Line chart on your worksheet using Ribbon options. **Insert tab → Charts group → Line drop-down menu → select "Line with Markers" style** (indicated in Figure 4.20).

The basic Line chart is shown in Figure 4.21.

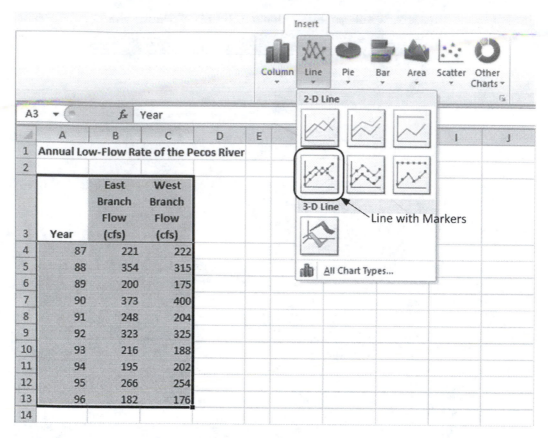

**Figure 4.20**
Inserting a Line chart with "Line with Markers" style.

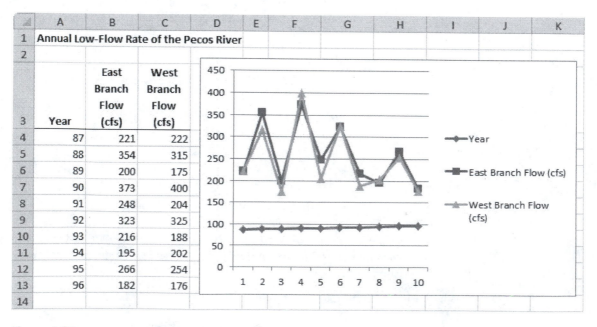

**Figure 4.21**
The basic Line chart (not quite what we wanted).

Notice (in Figure 4.21) that we did not get quite the chart we wanted. We wanted two data series plotted with the years on the *x*-axis as labels. Instead, Excel plotted the years as a third curve. It's easy to fix. First, select the "Year" curve on the plot and press the **Delete** key to remove the extra curve. The result is shown in Figure 4.22.

**Figure 4.22**

The Line chart after removing the "Year" curve.

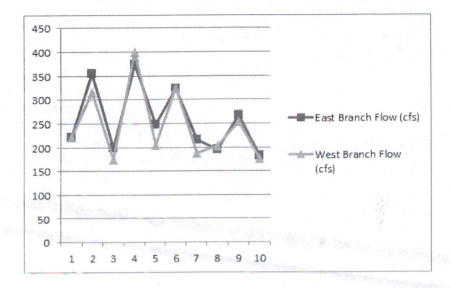

Next, we need to get the year values on the *x*-axis; follow these steps:

1. Click on the chart to select it. This causes the Chart Tools tabs to be displayed on the Ribbon.
2. On the Design tab, click the Select Data button: **Chart Tools → Design tab → Data group → Select Data button**. This opens the Select Data Source dialog shown in Figure 4.23.

**Figure 4.23**

The Select Data Source dialog box.

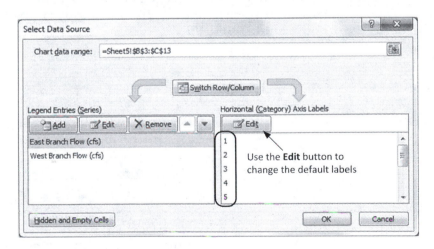

In Figure 4.23, the default labels (1,2,3,4,5, . . .) have been indicated. We need to change those to the values in cells A4:A13.

3. Click the **Edit** button in the Horizontal (Category) Axis Labels panel to change the *x*-axis labels.

4. Select cells A4:A13 as illustrated in Figure 4.24.
5. Click **OK** to close the Axis Labels dialog and return to the Select Data Source dialog.
6. Click **OK** to close the Select Data Source dialog.

**Figure 4.24**
Selecting the year labels for the *x*-axis.

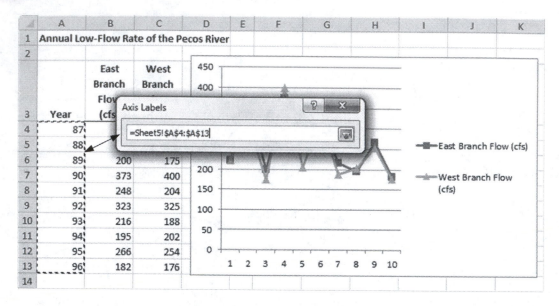

A few more changes are appropriate:

- Select a Chart Layout (Layout 1 was used): **Chart Tools → Design tab → Chart Layouts group → Layout 1.**
- Add text for the chart title and *y*-axis title.
- Add an *x*-axis title: **Chart Tools → Layout tab → Labels group → Axis Titles drop-down menu → Primary Horizontal Axis Title → Title Below Axis button.**
- Add text to the *x*-axis title.

The completed Line Graph is shown in Figure 4.25.

**Figure 4.25**
The completed Line chart.

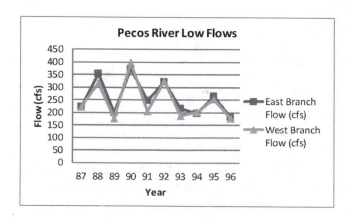

## PRACTICE!

To observe the difference between an XY Scatter chart and a Line chart, create both types of charts using the data set shown in Figure 4.26. The non-uniformly spaced *x* values will show the difference between XY Scatter charts and Line charts.

**Figure 4.26**
*XY* data with non-uniformly spaced *x* values.

|  | A | B | C |
|---|---|---|---|
| 1 | **X** | **Y** | |
| 2 | 1 | 1 | |
| 3 | 2 | 2 | |
| 4 | 8 | 3 | |
| 5 | 9 | 4 | |
| 6 | | | |

1. Create the worksheet shown in Figure 4.26.
2. Create an XY Scatter chart of the data. Your result should look like the left chart in Figure 4.27.
3. Create a Line chart of the same data. Your result should look like the right chart in Figure 4.27.

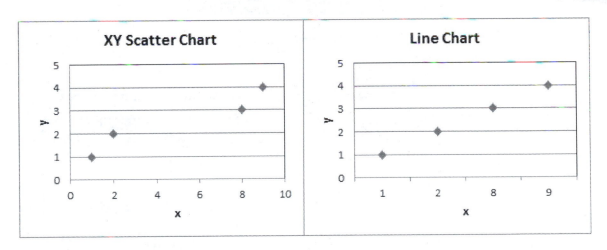

**Figure 4.27**
The same data, plotted with two different chart types.

Notice that, whenever you have non-uniformly spaced *x* values, the XY Scatter and Line chart types produce very different results. This is because the *x* values are considered numbers in the XY Scatter chart and category labels in the Line chart. That is why, on the Line chart, the *x*-axis shows 1, 2, 8, and 9 as labels centered between the ***tick marks*** on the *x*-axis.

Whenever you have numeric *x* values, you should use the XY Scatter chart type to correctly plot your data.

## PROFESSIONAL SUCCESS—FORMATTING CHARTS

The appearance of a chart in a document is important. A chart can make a lasting, visual impression that summarizes or exemplifies the main points of your presentation or document. The following formatting guidelines will help you create a professional-looking chart:

- A chart title should contain a clear, concise description of the chart contents.
- Create a label for each axis that contains, at a minimum, the name of the variable and the units of measurement that were used.
- Create a label for each data series. The labels can be consolidated in a legend if each data series is represented by a distinct color or style.
- Scale graduations should be included for each axis. The graduation marks may take the form of gridlines (uniformly spaced horizontal and vertical lines) or tick marks. The choice of scale graduations can be controlled from the Axis Options panel on the Format Axis dialog box.
- Ideally, scale graduations should follow the *1, 2, 5 rule*. The 1, 2, 5 rule states that one should select scale graduations so that the smallest division of the axis is a positive or negative integer power of 10 times 1, 2, or 5. For example, a scale graduation of 0.33 does not follow the rule.

## PRACTICE!

Practice formatting graphs for different purposes.

**Figure 4.28**
Linear and exponential growth curves—not formatted for a black-and-white text.

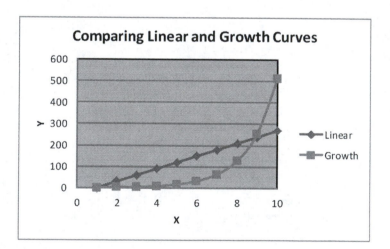

The chart shown in Figure 4.28 was created from the data shown in Figure 4.29. The data set is easy to generate; just enter the first two values in each column and then use the Fill Handle to complete the series. You will need to use the right mouse button with the Fill Handle to create the exponential growth series.

**Figure 4.29**

Data for linear and
exponential growth curves.

|  | A | B | C | D |
|---|---|---|---|---|
| 1 | Linear and Exponential Growth Curves | | | |
| 2 | | | | |
| 3 | X | Linear | Growth | |
| 4 | 1 | 0 | 1 | |
| 5 | 2 | 30 | 2 | |
| 6 | 3 | 60 | 4 | |
| 7 | 4 | 90 | 8 | |
| 8 | 5 | 120 | 16 | |
| 9 | 6 | 150 | 32 | |
| 10 | 7 | 180 | 64 | |
| 11 | 8 | 210 | 128 | |
| 12 | 9 | 240 | 256 | |
| 13 | 10 | 270 | 512 | |
| 14 | | | | |

First, create the XY Scatter chart from the data. Then format the chart for two different purposes:

1. Format the chart to be readable using a black-and-white printer.
   - Use distinct markers for each curve.
   - Use black for both curves, but change the line styles to make the two curves clearly distinct.
   - Set the plot area color to white. (This is important for black-and-white printing because it allows the chart to be photocopied without the plot area covering up the curves.)

2. Format the chart to use only color to distinguish the curves.
   - Use no markers for either curve.
   - Use solid lines for both curves.
   - Choose a bright color for the plot area.
   - Choose two colors for the curves that clearly differentiate the two curves on the colored background. (In this example, a medium background was used so that the line colors could be lighter and darker than the background so that the curves can be distinguished even when printed in black and white.)

Your results should look something like Figures 4.30 and 31.

**Figure 4.30**

The XY Scatter
chart formatted
for black-and-
white printer.

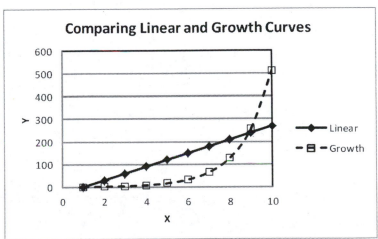

**Figure 4.31**
The XY Scatter chart formatted for color presentation.

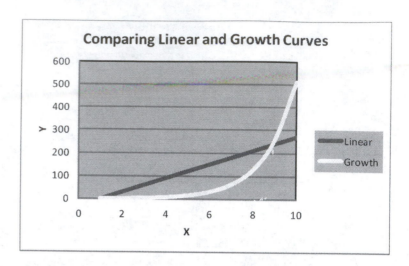

### 4.5.1 Changing Chart Types

The *chart type* can be changed after a chart has been created. Not all types of charts are appropriate for all data sets. For example, a pie chart is not appropriate for the Pecos River flow data used in this chapter (Table 4.2). But a *Column chart* will work with the data. To change the chart type from Line chart (shown in Figure 4.25) to Column chart, do the following:

1. Click on the chart to select it.
2. Click on the Change Chart Type button on the Chart Tools: Design tab. (Ribbon options. **Chart Tools → Design tab → Type group → Change Chart Type button.**) The Change Chart Type dialog box (Figure 4.32) will be displayed.
3. Select a Column chart style from the Column options. In this example, the "Clustered Column" style was selected.
4. Click **OK** to close the Change Chart Type dialog box.

**Figure 4.32**
The Change Chart Type dialog box.

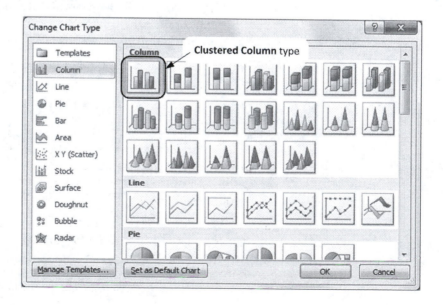

The result is shown in Figure 4.33.

**Figure 4.33**
The Pecos River flow data as a Column chart.

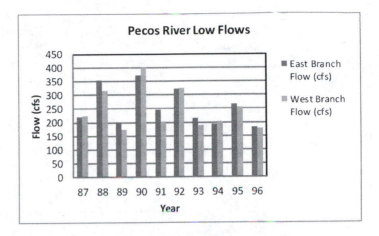

## 4.6 PREVIEWING AND PRINTING CHARTS

You can print charts and worksheets without previewing them, but using the print preview feature allows you to control how the chart or worksheet will appear when printed.

### 4.6.1 Previewing Charts

There are two ways to preview charts:

- Preview and Print the Worksheet with the Embedded Chart

A chart is typically embedded in a worksheet, and if you preview or print the worksheet, the chart will also be visible.

- Preview and Print the Chart Only

If you select an *embedded chart* before previewing or printing, then Excel will preview or print the chart only, without the rest of the worksheet. (A chart that was created as a separate page in the workbook will always preview and print separately.)

To preview a chart (only) before printing, do the following:

1. Click on the chart to select it.
2. Click the Print Preview button under the Office button: **File tab → Print tab**. (Printer controls and Print Preview are combined in Excel 2010.)

**Excel 2007**

1. Click on the chart to select it.
2. Click the Print Preview button under the Office button: **Office button → Print sub-menu → Print Preview button**. (Printer controls and Print Preview are handled separately in Excel 2007.)

In Print Preview the chart will be displayed as it will be printed. By default, when a chart is previewed or printed without the rest of the worksheet, Excel scales the chart to fit the page.

For more control over printing options, use the **Page Setup** link (shown in Figure 4.34) to open the Page Setup dialog, shown in Figure 4.35.

**Figure 4.34**
Previewing a selected chart.

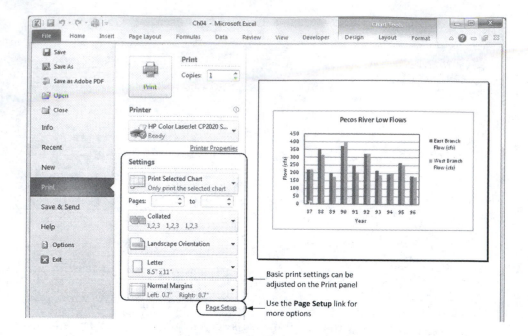

Basic print settings can be adjusted on the Print panel

Use the **Page Setup** link for more options

**Figure 4.35**
The Page Setup dialog box.

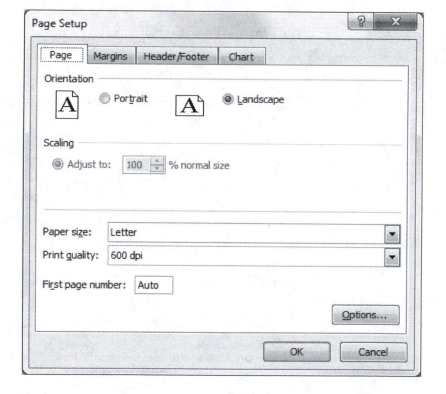

## 4.6.2 Printing Charts

You can print a chart by printing the worksheet that contains the chart, or you can print just the chart. To print the chart only, simply select the chart before printing.

When you have finished adjusting printing options, click **OK** on the Page Setup dialog (if it is open). Then clicking the **Print** button sends the chart to the printer.

## 4.7 ADDING DATA TO CHARTS

When working with plotted data, it is common to need to modify an existing chart to add additional data. You might have simply forgotten to type in a bit of data when creating the worksheet, or you might have collected additional data and need to include the new data on the existing chart. Excel makes either type of data addition easy.

### 4.7.1 Adding Data to an Existing Series

If you missed a row while typing in the data set, you might need to go back and add the missing data to the worksheet. When this happens, Excel makes it easy to update the chart as well. Procedurally, it makes a difference whether the missing data are being added to the middle of the data set or at the end, so each situation will be presented separately.

#### 4.7.1.1 Adding Data in the Middle of the Data Set

Suppose there was another data point in the Current versus Potential data set (Figure 4.1) that was accidentally overlooked when the data set was entered on the worksheet. The missing item is 0.022 A at 2.47 V.

To add the data in the middle of the data range, simply make some space for the data by moving the lower data values down one row or inserting a new row. The result is shown in Figure 4.36.

**Figure 4.36**
Making space for the additional data.

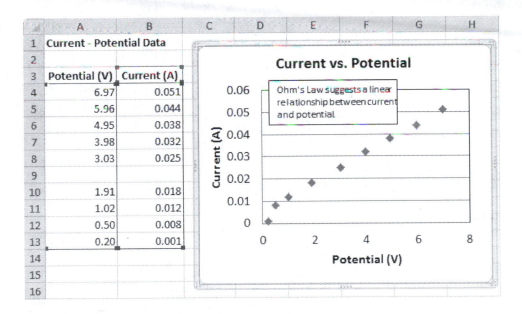

In Figure 4.36, the data series has been selected (by clicking on any of the markers). When the data series is selected, the data used to create the series are outlined by colored boxes on the worksheet. So, the box around cells A4:A13 indicates the *x* values used to create the data series, and the box around cells B4:B13 shows the *y* values used in the series. Notice that the empty cells created to hold the additional data are enclosed by the colored boxes that show the data used to create the data series. When you insert values in the middle of a data set, Excel assumes you want the data included in the chart. Adding the data to the worksheet in the middle of the data set will cause the new values to be included in the chart—automatically.

When the new values are added to the worksheet, the new point appears on the chart, as shown in Figure 4.37.

**Figure 4.37**
The XY Scatter chart with the new point indicated.

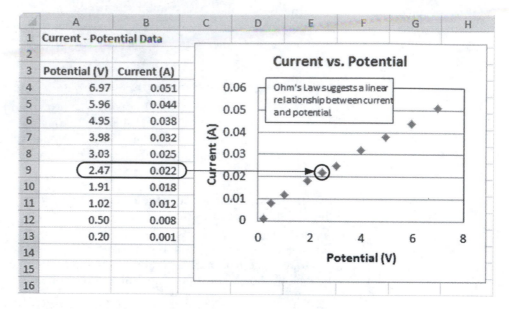

### 4.7.1.2 Adding Data at the End of the Data Set

When you add additional data at the end of the data set, Excel does not automatically include the new data in the plotted data set. For example, suppose that the experiment at the lowest potential was repeated because the data point at the left end of the data series seems a little out of line and we are concerned that there might have been a data collection error. When we repeat the experiment, we get 0.150 V at a current of 0.001 A. We do not want to replace the last data value, but we want to add the new data to the plot.

1. Add the new data to the worksheet in cells A14 and B14, as shown in Figure 4.38.

**Figure 4.38**
Data added at the end of the data set are not automatically plotted.

2. Click on any marker on the plotted data series to select the data series, as illustrated in Figure 4.38. Notice that the new data values in cells A14 and B14 are not included in the boxes that indicate the plotted data.

3. Position the mouse over either of the squares at the bottom edge of the Potential data box (the squares are called *handles*) and stretch the data box to include the new data value in cell A14. When the mouse is correctly positioned over the handle, the mouse icon changes to a double-headed arrow.

4. Repeat Step 3 to stretch the Current data box to include the new value in cell B14. The additional data point will be included on the Chart as part of the plotted data series.

### 4.7.1.3 Adding a Data Point to the End of a Data Set by Using the *Select Data Source Dialog Box*

If the chart exists in its own sheet (not embedded in a worksheet), then the colored boxes that indicate the data used to create a data series are not available. However, any data added in the middle of the data set used to create the chart will still be added to the chart automatically. But, if you need to add data to the end of the data set, you will need to use the Select Data Source dialog box, which is available using the **Select Data** button on the Ribbon: **Chart Tools → Design tab → Data group → Select Data button**.

The process for modifying the plotted data is as follows:

1. Click on the chart to select it. This causes the Ribbon to display the Chart Tools tabs. (This step is not needed if the chart is on its own page in the workbook; accessing the page selects the chart.)

2. Click the **Select Data** button on the Ribbon: **Chart Tools → Design tab → Data group → Select Data button**. This opens the Select Data Source dialog box shown in Figure 4.39.

3. Select the data series to be modified ("Current (A)" in this example).

4. Click the **Edit** button (indicated in Figure 4.39). This opens the Edit Series dialog box shown in Figure 4.40.

5. Select the cell ranges containing the data to be plotted.

6. Click **OK** to close the Edit Series dialog box.

7. Click **OK** to close the Select Data Source dialog box.

**Figure 4.39**
The Select Data Source dialog box.

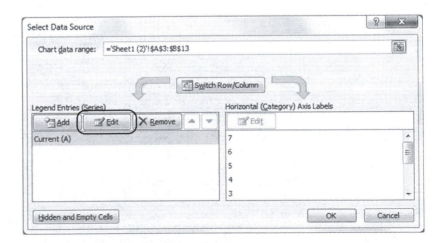

**Figure 4.40**

The Edit Series dialog box.

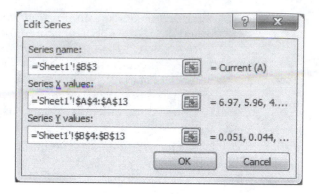

### 4.7.2 Adding a Data Series to an XY Scatter Chart

On a chart, each curve is a separate data series. On an XY Scatter chart, two data series may have differing *y* values but share *x* values; or there may be two sets of *x* values and two sets of *y* values. The procedure for adding a data series is different for each situation, so each will be presented separately.

#### 4.7.2.1 Adding a Second Data Series—Both Series Use the Same x Values

Repeating an experiment to test reproducibility is very common. For example, the experiment that generated the potential-current data shown in Figure 4.1 might be repeated. If it is possible to set exactly the same potential values across the resistor a second time, we could measure a new set of current values and obtain a data set containing one column of *x* values and two columns of *y* values. When plotted, the differences between the results from the two experiments would show how reproducible the experiment is (or isn't).

Repeating the experiment might generate a data set like that in Figure 4.41. In this figure, the chart has been selected to illustrate that the data from the second experiment has not yet been included on the chart.

**Figure 4.41**

Preparing to add data from the second experiment to the chart.

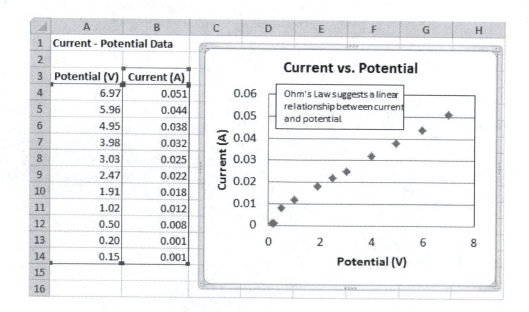

In the new data set, shown in Figure 4.41, there are two columns of *y* values (columns B and C), and one column of *x* values (column A). If you have already created a chart of the results from the first experiment, adding the *y* values from the second experiment is very easy.

1. Select the chart. This causes the data used to create the chart to be shown in boxes on the worksheet (Figure 4.41).
2. Use the mouse to stretch the box containing the *y* values in column B to include the new *y* values in column C. This is illustrated in Figure 4.42. (The marker sizes were reduced so that the data points are more distinct.)

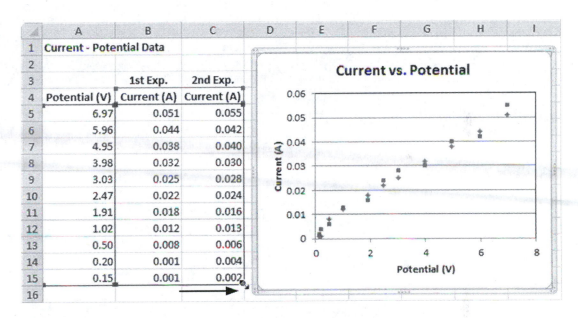

**Figure 4.42**
Expanding the *y* values cell range to include the second data series.

That's all it takes to add a new data series to a chart when the two data series use the same *x* values.

### 4.7.2.2 Adding a Second Data Series—Each Series Has Its Own x Values

If it is not possible to exactly reproduce the potentials, we can still repeat the experiment over the same range of potentials (approximately 0 to 7 V). If we did so, we would create a new data set with new *x* and *y* values, as shown in Figure 4.43.

If the new *y* values come with a new set of *x* values, then we can't just stretch the data boxes to create a second data series on the chart; we need to use the Select Data Source dialog box to add a new data series.

1. Click on the chart to select it. This causes the Ribbon to display the Chart Tools tabs.
2. Click the **Select Data** button on the Ribbon: **Chart Tools → Design tab → Data group → Select Data button**. This opens the Select Data Source dialog box shown in Figure 4.44.

**Figure 4.43**
Preparing to plot the data from the second experiment.

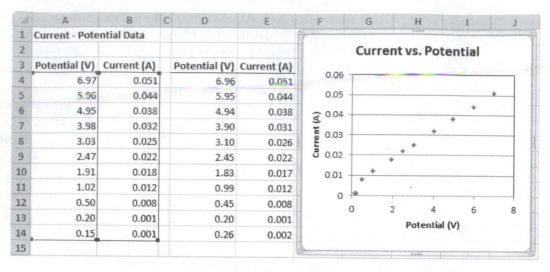

**Figure 4.44**
The Select Data Source dialog box used to add an additional data series to a chart.

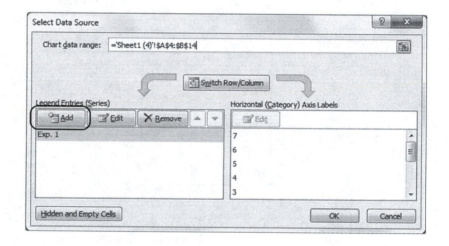

3. Click the **Add** button to add a new data series to the chart. This opens the Edit Series dialog box.
4. Select the cell ranges containing the new data to be plotted, cells D4:D14 and E4:E14 in this example (see Figure 4.45).

**Figure 4.45**
The *x* and *y* data values for the second curve are selected.

5. Click **OK** to close the Edit Series dialog box.
6. Click **OK** to close the Select Data Source dialog box.

The result is shown in Figure 4.46.

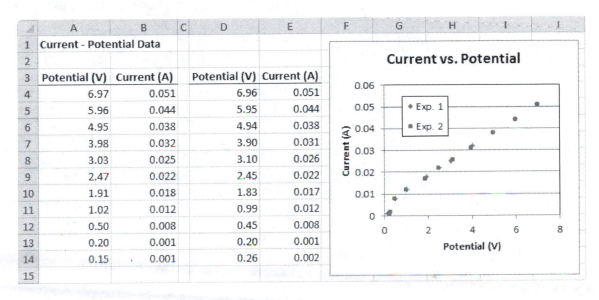

**Figure 4.46**
The XY Scatter chart with the new data series plotted.

## GRAPHING TO EVALUATE A FUNCTION

APPLICATION

Visualization can be a big help in trying to understand what your data means or how a function works. Excel can help with this by allowing you to quickly and easily graph data by evaluating a function. For example, exponentials and hyperbolic sines are commonly used functions for solving differential equations. When solving these equations, you select the appropriate function on the basis of its characteristics. Being able to see a graph of a function is a big help in understanding how the function behaves. Figure 4.47 shows a plot of the hyperbolic sine function $(SINH(x))$ from $x=-10$ to 10.

**Figure 4.47**
Plotting the hyperbolic sine function ($SINH$).

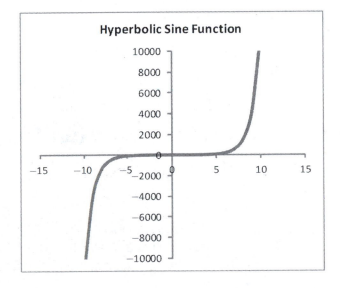

**How Did They Do That?**

To help visualize a function, first evaluate the function over a range of values in a worksheet, as shown in Figure 4.48.

**Figure 4.48**
Evaluation of the hyperbolic sine function, *SINH*.

| | A | B | C | D |
|---|---|---|---|---|
| | B4 | ▾ | $f_x$ =SINH(A4) | |
| 1 | Hyperbolic Sine | | | |
| 2 | | | | |
| 3 | x | sinh(x) | | |
| 4 | -10 | -11013 | | |
| 5 | -9 | -4052 | | |
| 6 | -8 | -1490 | | |
| 7 | -7 | -548 | | |
| 8 | -6 | -202 | | |
| 9 | -5 | -74 | | |
| 10 | -4 | -27 | | |
| 11 | -3 | -10 | | |
| 12 | -2 | -4 | | |
| 13 | -1 | -1 | | |
| 14 | 0 | 0 | | |
| 15 | 1 | 1 | | |
| 16 | 2 | 4 | | |
| 17 | 3 | 10 | | |
| 18 | 4 | 27 | | |
| 19 | 5 | 74 | | |
| 20 | 6 | 202 | | |
| 21 | 7 | 548 | | |
| 22 | 8 | 1490 | | |
| 23 | 9 | 4052 | | |
| 24 | 10 | 11013 | | |
| 25 | | | | |

Then, create an XY Scatter plot of the evaluated results, select a chart layout, and enter the chart title. Set the axis formatting as desired.

## 4.8 CHARTING FEATURES USEFUL TO ENGINEERS

Excel has many advanced features for formatting charts. Some of the features are particularly useful for engineering applications. These include:

- Trendlines
- Error bars
- *Logarithmic axes*
- Secondary axes

### 4.8.1 Adding a Trendline to a Chart

A *trendline* is a best-fit line through a set of data points. Excel makes it very easy to add a trendline to an XY Scatter chart, and will present the equation of the trendline on the chart. Trendlines are available from the Chart Tools: Layout tab. (Ribbon options. **Chart Tools → Layout tab → Analysis group → Trendline drop-down menu**).

**Figure 4.49**
Trendlines are available from the Chart Tools: Layout tab.

Excel provides several commonly used trendline options, but only two are available from the Trendline drop-down menu on the Ribbon (indicated in Figure 4.49).

- Linear (straight line through the data)
- Exponential

Use the **More Trendline Options . . .** button to open the Trendline dialog box for additional trendline options (shown in Figure 4.50). The Trendline dialog box is also used to tell Excel to display the equation of the trendline on the chart.

**Figure 4.50**
The Trendline drop-down menu provides quick access to commonly used trendlines.

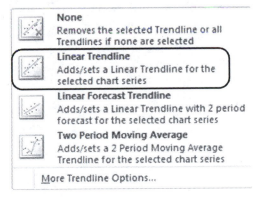

Because Ohm's law predicts a linear relationship between potential and current, we will add a linear trendline to the potential-current data. To add a linear trendline, follow these steps:

1. Click on any marker to select the data series.
2. Select Linear Trendline from the Trendline drop-down menu: **Chart Tools → Layout tab → Analysis group → Trendline drop-down menu → Linear Trendline option**.

Excel will show the trendline on the chart, as shown in Figure 4.51.

**Figure 4.51**
The current versus potential data with a linear trendline added.

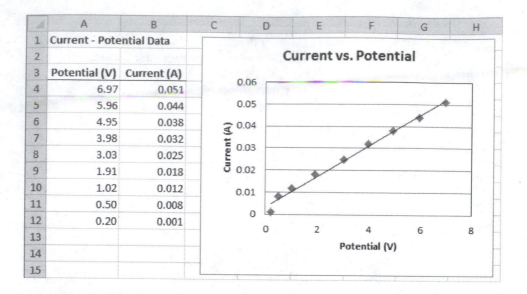

By default Excel does not show the equation of the trendline. To see the equation:

1. Right-click on the trendline (stay away from the markers) and select **Format Trendline . . .** from the pop-up menu as illustrated in Figure 4.52 (or use Ribbon options. **Chart Tools → Layout tab → Analysis group → Trendline drop-down menu → More Trendline Options . . .**). This opens the Format Trendline dialog box, shown in Figure 4.53.
2. Check the **Display Equation on chart** box.

The result is shown in Figure 4.54. Other commonly used options, indicated in Figure 4.53, allow you to:

- Set the intercept of the trendline to a particular value (frequently zero).
- Display the $R^2$ value for the trendline on the chart. An $R^2$ value of 1 implies a perfect fit to the data. The lower the $R^2$ value, the poorer the fit.

**Figure 4.52**
Select Format Trendline . . . to open the Format Trendline dialog box.

| | A | B | C | D | E | F | G | H |
|---|---|---|---|---|---|---|---|---|
| 1 | Current - Potential Data | | | | | | | |
| 2 | | | | | | | | |
| 3 | Potential (V) | Current (A) | | | | | | |
| 4 | 6.97 | 0.051 | | | | | | |
| 5 | 5.96 | 0.044 | | | | | | |
| 6 | 4.95 | 0.038 | | | | | | |
| 7 | 3.98 | 0.032 | | | | | | |
| 8 | 3.03 | 0.025 | | | | | | |
| 9 | 1.91 | 0.018 | | | | | | |
| 10 | 1.02 | 0.012 | | | | | | |
| 11 | 0.50 | 0.008 | | | | | | |
| 12 | 0.20 | 0.001 | | | | | | |
| 13 | | | | | | | | |
| 14 | | | | | | | | |
| 15 | | | | | | | | |

**Current vs. Potential**

Delete
Reset to Match Style
Change Chart Type...
Select Data...
3-D Rotation...
Format Trendline...

**Figure 4.53**
The Format Trendline dialog box.

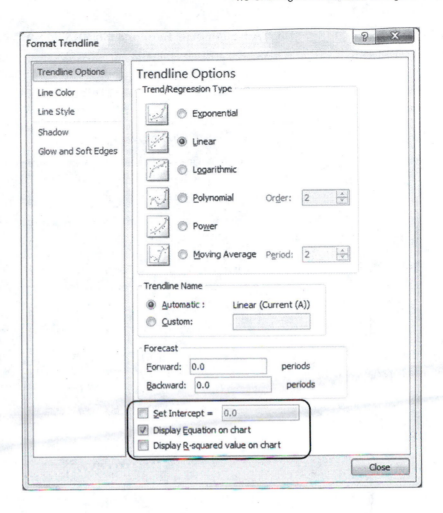

**Figure 4.54**
The XY Scatter chart with trendline and equation displayed.

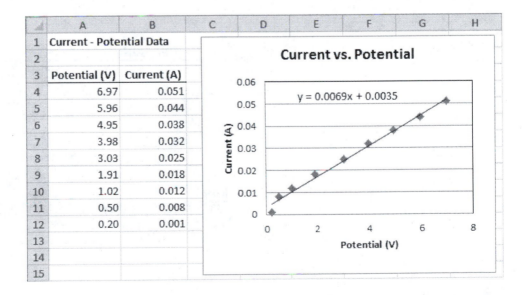

### 4.8.2 Adding Error Bars to a Chart

*Error bars* represent the range of measured or statistical error in a data series. Error bars should not be used unless you understand their purpose.

The flow rates used in preparing the Annual Low Flows for Pecos River chart are only accurate to ±5% of each data value. We will use error bars to indicate this level of uncertainty. To add error bars to the East Branch data series shown in Figure 4.33, follow this procedure:

1. Select the series by clicking on any column of East Branch data.
2. Use the Error Bars drop-down menu from the Ribbon: **Chart Tools → Layout tab → Analysis group → Error Bars drop-down menu.**

The Error Bars drop-down menu contains various options as shown in Figure 4.55. One of the options is error bars with a 5% value (selected in Figure 4.55), so we can add the error bars we need with one click. Usually you will need to use the **More Error Bars Options . . .** button to open the Error Bars dialog box.

The Column chart with 5% error bars is shown in Figure 4.56.

**Figure 4.55**

The Error Bars drop-down menu.

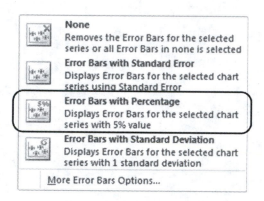

**Figure 4.56**

The Column chart with 5% error bars.

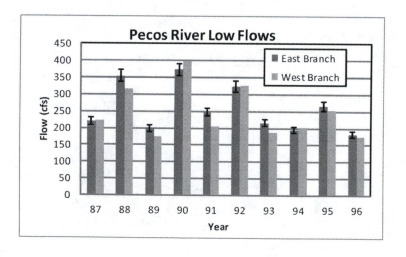

### 4.8.3 Using a Log Scale Axis

Excel will allow you to select a *logarithmic scale* on either the *x*- or *y*-axis as long as taking a logarithm of the values plotted on the axis is valid (no zero or negative values). If the exponential growth curve shown in Figure 4.28 is plotted with a logarithmic *y*-axis, the data series should plot as a straight line.

To instruct Excel to use a log scale on the *y*-axis, do the following:

1. Click on any number along the *y*-axis to select the axis. This causes the Chart Tools tabs on the Ribbon to be displayed.
2. Use the Format Selection button on the Layout (or Format) tab to open the Format Axis dialog box (Figure 4.57). Ribbon options: **Chart Tools → Layout tab → Current Selection group → Format Selection button**.
3. Check the **Logarithmic scale** box, indicated in Figure 4.57.
4. Click **Close** to close the Format Axis dialog box.

The result is shown in Figure 4.58.

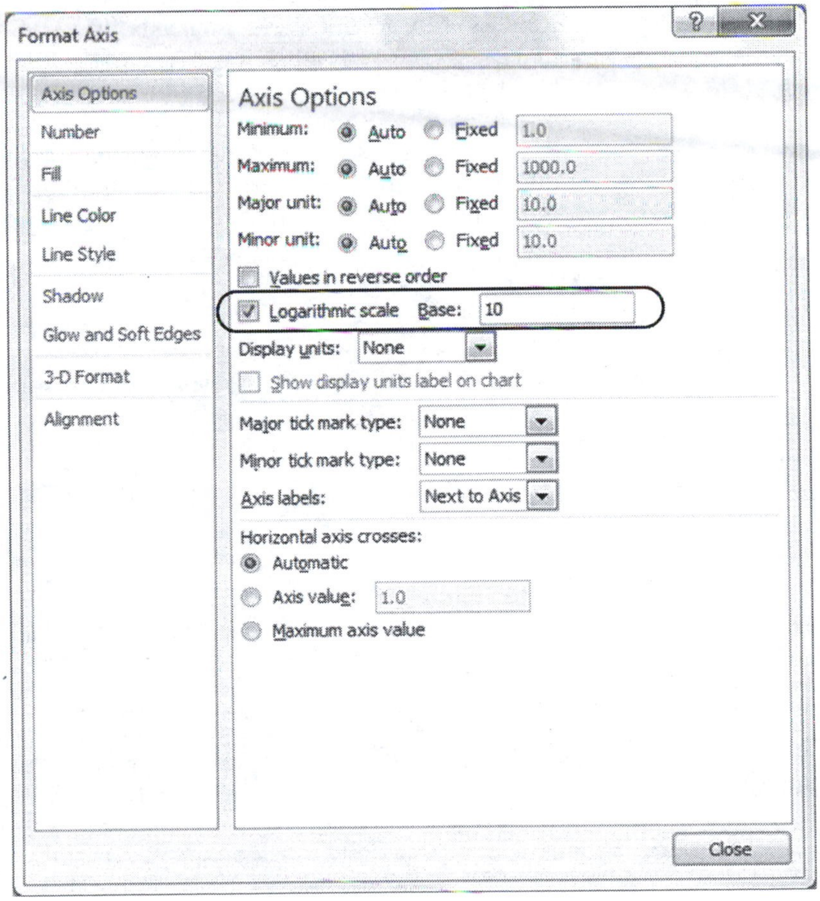

**Figure 4.57**
The Format Axis dialog box.

**Figure 4.58**
Exponential growth, plotted
on a log scale.

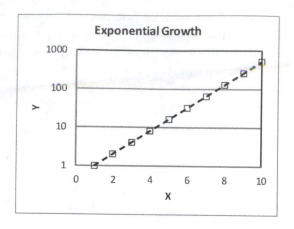

## 4.8.4 Using Secondary Axes

A *secondary axis* is a second *y*-axis on the right side of the chart. It is typically used when two related but distinct data series are plotted on the same chart. That is, the two data series are both dependent upon the variable plotted on the *x*-axis, but have widely varying *y* values, and usually different units. By plotting one series on each of the *y* axes (left and right), it can be easier to understand the chart.

To see how two related but different data series can be plotted on the same graph, consider the (fictional) data set shown in Figure 4.59, which shows the low flows in the East Branch of the Pecos River and the average daily high temperature during the summer months. Presumably, there should be a relationship between a hot summer and low water.

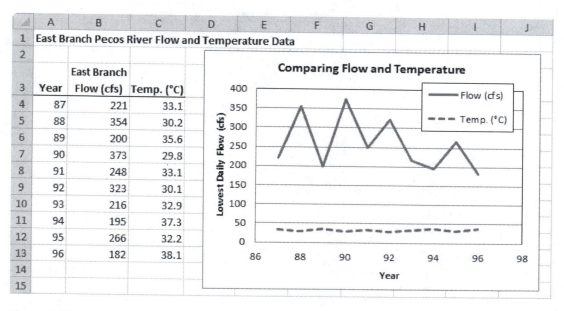

**Figure 4.59**
Pecos River flow and temperature data, before using a secondary axis.

Because the temperature values are ten times smaller than the flow values, when they are plotted on the same scale, as in Figure 4.59, the temperature curve looks like a fairly flat line at the bottom of the chart. To see more detail in the temperature curve, we will plot it using a secondary axis. To request a secondary axis, follow these steps:

1. Click on the temperature curve to select the data series.
2. Click the Format Selection button on the Layout tab: **Chart Tools → Layout** (or **Format**) **tab → Current Selection group → Format Selection button**. This opens the Format Series dialog shown in Figure 4.60.
3. Select the **Secondary Axis** option in the **Plot Series On** list.
4. Click **Close** to close the Format Series dialog box.

The chart with the secondary axis is shown in Figure 4.61.

**Figure 4.60**
The Format Data Series dialog box.

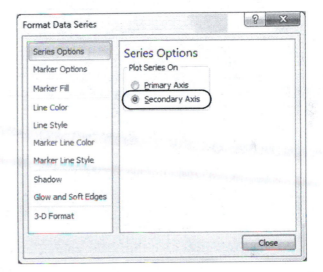

**Figure 4.61**
The Pecos River flow and temperature data, using a secondary axis.

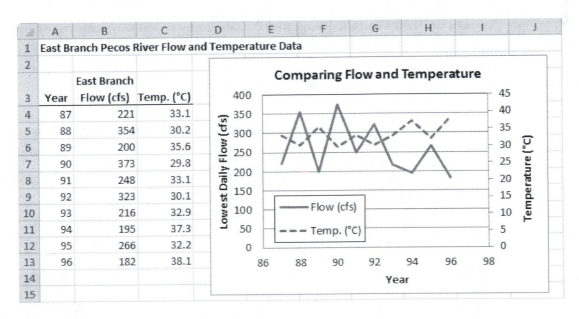

## KEY TERMS

| | | |
|---|---|---|
| 1, 2, 5 rule | gridlines | secondary axis |
| axis labels | handles | series (data series) |
| chart | labels | smoothed lines |
| chart layout | legend | text box |
| chart type | Line chart | tick mark |
| Column chart | lines | trendline |
| data series | logarithmic axis | $x$-axis |
| embedded chart | markers | XY Scatter chart |
| error bars | plot area | $y$-axis |

## SUMMARY

**Creating an XY Scatter Chart**

1. Select the data to be plotted.
2. Use the Ribbon's **Insert tab → Charts Group → Scatter drop-down menu**.
3. Select the type of chart to be created from the Scatter drop-down menu.
4. Select a Chart Layout: **Chart Tools → Design tab → Chart Layouts group → Chart Layouts palette**.
5. Format individual chart features as needed.

**Creating a Line Chart**

Do not use a Line chart when your data includes numeric, non-uniformly spaced $x$ values; use an XY Scatter plot instead.

1. Select the data to be plotted.
2. Use the Ribbon's **Insert tab → Charts Group → Line drop-down menu**.
3. Select the type of chart to be created from the Line drop-down menu.
4. Select a Chart Layout: **Chart Tools → Design tab → Chart Layouts group → Chart Layouts palette**.
5. Format individual chart features as needed.

**Annotating a Chart by Adding a Text Box**

1. Select the chart.
2. Click the **Text Box** button: **Chart Tools → Layout tab → Insert group → Text Box button**.
3. Position the mouse cursor on the chart where you want to place the text and then click the mouse.
4. Type the desired text.
5. Click outside the text box to complete the text entry.

### Using a Dialog Box to Format a Chart Feature

1. Select the chart feature by clicking on it.
2. Click the **Format Selection** button on the Ribbon: **Chart Tools → Layout tab → Current Selection group → Format Selection button**. This opens the formatting dialog box for the chart feature.
3. Set the desired format options.
4. Click the **Close** button when you are done making changes.

### Changing the Chart Type

1. Click on the chart to select it.
2. Click on the Change Chart Type: **Chart Tools → Design tab → Type group → Change Chart Type button**.
3. Select a chart style.
4. Click **OK** to close the Change Chart Type dialog box.

### Previewing and Printing Charts

- Preview and Print the Worksheet with the Embedded Chart
- **Preview and Print the Chart Only** (Select chart before previewing or printing)

*Excel 2010*

**File tab → Print tab**

*Excel 2007*

Previewing: **Office button → Print sub-menu → Print Preview button**
Printing: **Office button → Print sub-menu → Print Preview button**

### Adding Data to Charts

*Adding Data to an Existing Series—Data in the Middle of the Data Set*

1. Insert one or more rows in the middle of the data set.
2. Enter the new data values. The chart will automatically include the new data values.

*Adding Data to an Existing Series—Data at the End of the Data Set*

1. Insert one or more rows in the middle of the data set.
2. Enter the new data values. The chart will not automatically include the new data values.
3. Select the chart to cause the data plotted on the chart to be indicated (colored boxes).
4. Adjust the size of the colored boxes to include the new data values.

*Adding Data to an Existing Series Using the Select Data Source Dialog Box*

1. Click on the chart to select it.
2. Click the **Select Data** button: **Chart Tools → Design tab → Data group → Select Data button**.
3. Select the data series to be modified.
4. Click the Edit button.
5. Select the cell ranges containing the data to be plotted.

6. Click **OK** to close the Edit Series dialog box.

7. Click **OK** to close the Select Data Source dialog box.

*Adding a Second Data Series—Both Series Use the Same x Values*

1. Select the chart.

2. Use the mouse to adjust the size of data box to include the new *y* values.

*Adding a Second Data Series—Each Series Has Its Own x Values*

1. Click on the chart to select it.

2. Click the **Select Data** button: **Chart Tools → Design tab → Data group → Select Data button**.

3. Click the **Add** button.

4. Select the cell ranges containing the new data to be plotted.

5. Click **OK** to close the Edit Series dialog box.

6. Click **OK** to close the Select Data Source dialog box.

### Adding a Trendline

1. Right-click on a data series.

2. Select **Insert Trendline** from the pop-up menu.

3. Choose the Trendline type from the dialog box.

4. Set options as desired:
   (a) Set intercept value (if applicable).
   (b) Display equation of trendline on chart.
   (c) Display the $R^2$ value of the regressed trendline.

5. Click **Close** to close the dialog box.

### Adding Error Bars

1. Click on a data series.

2. Use the Error Bars menu: **Chart Tools → Layout tab → Analysis group → Error Bars drop-down menu**.

3. Choose the type of error bar desired from the drop-down menu, or select **More Error Bar Options . . .** to open the dialog box.

### Using a Log Scale Axis

1. Click on any number along the *y*-axis to select the axis.

2. Use the Format Selection button: **Chart Tools → Layout tab → Current Selection group → Format Selection button**.

3. Check the **Logarithmic scale** box.

4. Click **Close** to close the Format Axis dialog box.

### Using a Secondary Axis

1. Click on the data series that is to be plotted using the secondary axis.

2. Click the Format Selection button: **Chart Tools → Layout (or Format) tab → Current Selection group → Format Selection button**.

3. Select the **Secondary Axis** option in the **Plot Series On** list.

4. Click **Close** to close the Format Series dialog box.

# PROBLEMS

1. Generate data points for the function

$$y = 4\sin(x) - x^2$$

   For $x = -5.0, -4.5, ..., 4.5, 5.0$. Chart the results by using an XY scatter plot. Add appropriate title and axis labels.

2. The equation for the plot of a circle is

$$x^2 + y^2 = r^2,$$

   where $r$ is the radius of the circle. For a radius of 5, the Excel equation for $y$ is

$$y = \mathrm{SQRT}(5 - x{\char`\^}2)$$

   Generate points for $y$ for $x = 0.0, ..., 2.2$ in increments of 0.1, and plot the results. Why does your plot show only ¼ of a circle? Can you create more data points to plot a full circle?

3. The resolution of the data can dramatically change the appearance of a graph. If there are too few data points, the plot will not be smooth. If there are too many data points, the time and storage requirements become a burden. Generate two sets of data points for the following function:

$$f(x) = \sin(x)$$

   For the first set use $x = 0.1, ..., 25.0$ in increments of 0.1 (250 data points). (You will definitely want to use the Fill Handle for this!) Generate another set with $x = 0, ..., 25$ in increments of 1 (25 data points). Plot both versions of $\sin(x)$. Does the plot with 25 data points give you a correct impression of the shape of the sine function?

4. A graph that uses logarithmic scales on both axes is called a log–log graph. A log–log graph is useful for plotting power equations, since they appear as straight lines. A power equation has the following form:

$$y = ax^b$$

   Table 4.3 presents data collected from an experiment that measured the resistance of a conductor for a number of sizes. The size (cross-sectional

**Table 4.3 Resistance versus Area of a Conductor**

| Area (mm$^2$) | Resistance (mΩ/m) |
| --- | --- |
| 0.009 | 2000.0 |
| 0.021 | 1010.0 |
| 0.063 | 364.0 |
| 0.202 | 110.0 |
| 0.523 | 44.0 |
| 1.008 | 20.0 |
| 3.310 | 8.0 |
| 7.290 | 3.5 |
| 20.520 | 1.2 |

**Table 4.4 Average Daily Traffic Flow at Four Downtown Intersections**

| | Average Daily Traffic Flow (×1,000) | | | |
| --- | --- | --- | --- | --- |
| | Intersection # | | | |
| Year | 1 | 2 | 3 | 4 |
| 1996 | 25.3 | 12.2 | 34.8 | 45.3 |
| 1997 | 26.3 | 14.5 | 36.9 | 48.7 |
| 1998 | 28.6 | 14.9 | 42.6 | 43.2 |
| 1999 | 29.0 | 16.8 | 50.6 | 46.9 |
| 2000 | 32.4 | 17.6 | 70.8 | 54.9 |
| 2001 | 34.8 | 17.9 | 82.3 | 60.9 |

area) was measured in millimeters squared, and the resistance was measured in milliohms per meter. Create a scatter plot of these data.

5. Modify the *x* and *y* axes of the chart created in the previous problem to use a logarithmic scale. From viewing the resulting scatter plot, what can you infer about the relationship between resistance and size of a conductor in this experiment?

6. Table 4.4 shows the average daily traffic flow at four different intersections for a five-year period. Create a worksheet and enter the table. Plot the data for intersections 1, 3, and 4, but not for intersection 2. Use different line types and markers for each of the three lines. Your graph should look like Figure 4.62.

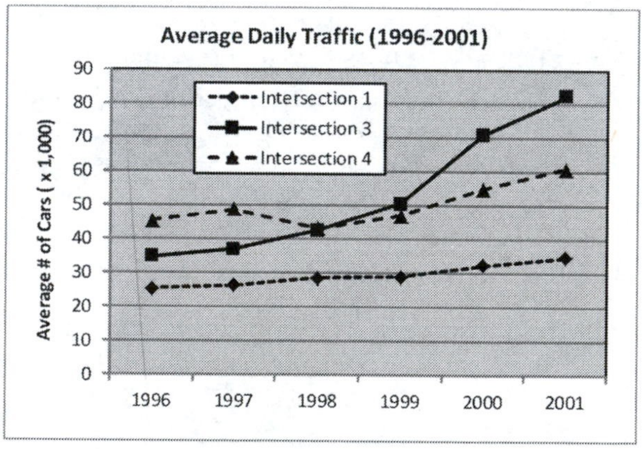

**Figure 4.62**
A plot of three of the columns in Table 4.4.

7. The Help system in Excel contains a topic entitled *Present your data in a scatter chart or a line chart.* Type the topic title into the Help system's search box, and then read the article. Summarize the differences between XY Scatter charts and Line charts.

8. You can use a chart to find the solution to an equation. For example, the function

$$f(x) = 4x^3 + 12x^2 - 64x + 16 = 0$$

should have three solutions because of the $x^3$. The solutions might not all be real, and could be repeated, but for this particular equation there are three distinct, real solutions, or *roots*. By simply evaluating the function over a range of *x* values and charting the results, approximate solutions can be seen on the graph. Figure 4.63 shows how this is done and indicates that $x = 0.25$ and $x = 2.6$ are approximate solutions.

**Figure 4.63**

Using a chart to find solutions to an equation. Solutions are indicated.

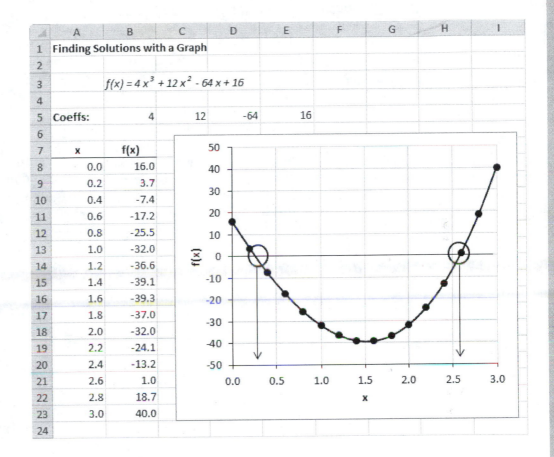

Modify the *x* values in column A to try to:

**a.** find the value of the third solution;

**b.** get more precise values for the solutions.

9. Create a chart to find approximate solutions of the following equation; then solve the quadratic equation to check your results:

$$3x^2 - 8x + 10 = 0$$

10. Use a chart to find the values of *x* between 0.1 and 3 that satisfy the following equation:

$$4\sin(x) + 12x^2 - 63\ln(x) + 16 = 0$$

11. Enthalpy and entropy data for saturated steam between 100 and 300°C are shown in Table 4.5.

**a.** Prepare a chart of the enthalpy data as a function of temperature, and use it to estimate the enthalpy of saturated steam at 135°C.

**b.** Prepare a chart of the entropy data as a function of temperature, and use it to estimate the entropy of saturated steam at 260°C.

**Table 4.5 Properties of Saturated Steam**

| Temp. (°C) | Enthalpy (kJ/kg) | Entropy (kJ/kg K) |
|---|---|---|
| 100 | 2,680 | 7.4 |
| 150 | 2,750 | 6.8 |
| 200 | 2,790 | 6.4 |
| 250 | 2,800 | 6.1 |
| 300 | 2,750 | 5.7 |

12. Finance majors are often adjusting cost and revenue equations to try to maximize profit, which is the difference between revenue and cost ($P = R - C$). The following cost and revenue equations are functions of $x$, the number of units produced and sold. Calculate the profit for a range of $x$ values, and create a chart of profit versus units produced. What value of $x$ produces the maximum profit? (Hint: It's less than 200.)

$$C = 1.2x^2 + 14x + 65$$
$$R = 527x - 0.015x^3$$

# 5

# Performing Data Analysis

## Sections

## Objectives

*After reading this chapter, you should be able to perform the following tasks:*

- Access and use the Analysis ToolPak.
- Create a histogram.
- Calculate descriptive statistics for a data series.
- Calculate the correlation between two data series.

- Perform a linear regression analysis on a set of data.
- Calculate linear and exponential trends for data series.
- Add trendlines to charts.
- Undertake the iterative solution of equations by using the Goal Seek.
- Perform optimization by using the Solver tool.

## 5.1 INTRODUCTION

Engineers are routinely asked to make decisions and recommendations on the basis of data sets. The data sets may come from laboratory experiments or from a manufacturing plant's quality assurance (QA) tests; either way, ***data analysis*** is the process used to get from raw data to the results that can be used to make decisions. Excel is commonly used for data analysis, in part because it provides a number of tools that can simplify the data analysis process. Some of those tools will be presented here, including those that perform the following functions:

- Creating a histogram from a data set.
- Calculating basic descriptive statistics about a data set.
- Checking for a correlation between two data sets.
- Performing a linear regression.
- Using a trend analysis to predict future values on the basis of historical values.

Also, engineers often use *iterative methods* to solve complex equations. When using an iterative method, the equation is solved over and over again, changing the values of the input variables each time until the equation is satisfied. For example, the equation

$$3x^3 + 4x^2 - 2x - 3 = 0$$

can be solved in several ways. One option is to use an iterative method to simply try various values of $x$ until you find the values that satisfy the equation. Two ways to solve problems using iterative methods in Excel are the *Goal Seek* and the *Solver*. The Solver can also be used to find the best solution; this process is called *optimization*.

## 5.2 USING THE ANALYSIS TOOLPAK

An add-in package is available for Excel that includes a number of statistical and engineering tools. This package, called the *Analysis ToolPak*, can be used to shorten the time that it usually takes to perform a complex analysis.

The Analysis ToolPak is an add-in for Excel that is typically installed as part of the Excel program, but not activated. Excel keeps program size down and performance up by not activating the add-ins that are not commonly used.

In Excel 2007 or 2010, a quick glance at the Ribbon's Data tab (Figure 5.1) will show you if the Analysis ToolPak is ready for use on your system. If a button labeled **Data Analysis** appears on the Data tab in the Analysis group, then the Analysis ToolPak is installed and active on your system. If the button is missing, the Analysis ToolPak must either be activated or (less commonly) be installed from the Excel program CDs.

**Figure 5.1**

If the Data Analysis button appears on the Ribbon's Data tab, the Analysis ToolPak is installed and active.

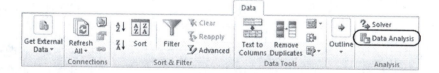

(In Excel 2003, Choose Tools from the Menu bar. If the Data Analysis command does not appear on the Tools menu, then the Analysis ToolPak must be either activated or installed.)

### 5.2.1 Activating the Analysis ToolPak

If the **Data Analysis** button appears on the Ribbon's Data tab, you can skip this step.

In Excel 2007 and 2010, add-ins like the Analysis ToolPak are managed using the Add-Ins panel on the Excel Options dialog box. To access the Excel Options dialog box, use the following options:

- Excel 2010: **File tab → Options button**
- Excel 2007: **Office button → Excel Options button**

The Excel Options dialog, shown in Figure 5.2, will open. Use the **Add-Ins** panel to activate the Analysis ToolPak.

**Figure 5.2**

The Add-Ins panel on the Excel Options dialog box.

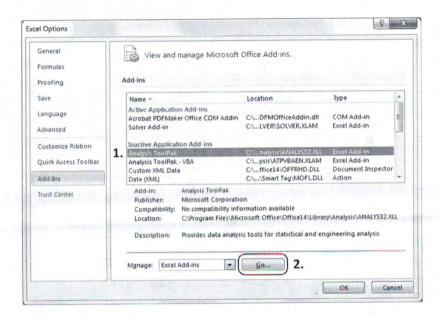

Look for "Analysis ToolPak" in the **Inactive Application Add-Ins** list (highlighted in Figure 5.2).

- If "Analysis ToolPak" is not listed in the **Inactive Application Add-Ins** list, it must be installed from the Excel program CDs (this is not common).
- If "Analysis ToolPak" is listed in the **Inactive Application Add-Ins** list, the Analysis ToolPak has been installed and you simply need to activate it.

The Analysis ToolPak can be activated with the following steps:

1. Select "Analysis ToolPak" in the **Inactive Application Add-Ins** list.
2. Click **Go . . .** to open the Add-Ins dialog box, shown in Figure 5.3.
3. Check the box labeled **Analysis ToolPak** (as illustrated in Figure 5.3).
4. Click **OK** to close the Add-Ins dialog box.
5. Click **OK** to close the Excel Options dialog box.

The Analysis ToolPak is now active on your system.

**Note:** In Excel 2003, open the Add-Ins dialog box using menu options: **Tools → Add-Ins**, then check the box labeled **Analysis ToolPak** as illustrated in Figure 5.3.

### 5.2.2 Opening the Data Analysis Dialog Box

Once the Analysis ToolPak is active, click the **Data Analysis** button on the Ribbon's Data tab (see Figure 5.1) to open the Data Analysis dialog box, shown in Figure 5.4. (Excel 2003: Use menu options Tools → Data Analysis.)

**Figure 5.3**
The Add-Ins dialog box.

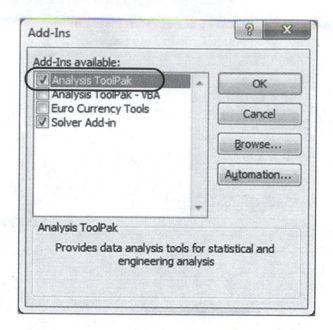

**Figure 5.4**
The Data Analysis dialog box.

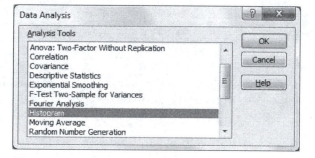

The Data Analysis dialog box provides access to a wide assortment of analysis tools. Each tool requires a set of input parameters in a specific format, and dialog boxes are used to collect the required pieces of information. These usually include:

- an input range (cells containing the data to be analyzed)
- an output range (a place for the tool to put the calculated results)
- option settings to control the data analysis process

The results of the data analysis are displayed in an output table. Additionally, some tools will generate a chart.

We will demonstrate how to use the Analysis ToolPak by using four of the available tools to perform these tasks:

- Create a histogram.
- Provide descriptive statistics.
- Compute a correlation.
- Perform a regression analysis.

It is beyond the scope of this text to interpret the results of these statistical analyses or to explain the meaning of statistical terms such as confidence interval, residuals, $R$ square, standard error, etc.

## 5.3 CREATING A HISTOGRAM

We will first demonstrate the use of the Analysis ToolPak with the creation of a *histogram*. A histogram is a graph of the frequency distribution of a set of data. That is, it is a chart that shows how often certain values appear in a data set. Most students have seen histograms used by a teacher to show how many As, Bs, and so on were given on a particular assignment. An example of a grade distribution chart, or grade histogram, is shown in Figure 5.5.

**Figure 5.5**
Grade histogram.

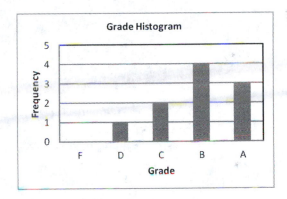

To create a histogram, the data are aggregated, or separated into groups, and the groups are graphed in a Column chart. A group is sometimes called a *bin*, since the process of creating a histogram can be visualized as follows:

1. Sort the data into bins.
2. Count the number of values in each bin.
3. Graph the bin names on the x-axis and the number of values in the bins on the y-axis.

In the grade example, each letter grade is assigned to a bin. To create the graph in Figure 5.1, we identify the bins (A to F), count the number of grades in each bin, and create the Column chart.

**Note:** The Histogram tool in Excel will only create histograms of numeric data, not the letter grades used in this example. In the next example we will use numeric scores and the Histogram tool to create a grade histogram.

In a histogram, each column represents a bin, and the height of the bar represents the number of data values in that bin, which is called the *frequency* of data in the bin.

## EXAMPLE 5.1

### CREATING A GRADE HISTOGRAM

We begin by looking at a data set containing the student scores, Figure 5.6.

**Figure 5.6**
Student score data.

| | A | B | C |
|---|---|---|---|
| 1 | Grade Histogram | | |
| 2 | | | |
| 3 | Student | Score | |
| 4 | Sal | 87 | |
| 5 | Sali | 94 | |
| 6 | Sally | 72 | |
| 7 | Sam | 82 | |
| 8 | Sara | 88 | |
| 9 | Sarah | 78 | |
| 10 | Su Nee | 95 | |
| 11 | Stewart | 91 | |
| 12 | Stuart | 68 | |
| 13 | Suri | 84 | |
| 14 | | | |

Next, we decide how the scores should be related to grades. A common grading method uses 10-point intervals for each grade level, or bin. Scores less than 60 receive a grade of F, less than 70 receive a D, and so on. This is indicated on the worksheet by defining bins as shown in column E in Figure 5.7.

**Figure 5.7**
Defining the upper limits of each grade bin.

| | A | B | C | D | E | F |
|---|---|---|---|---|---|---|
| 1 | Grade Histogram | | | | | |
| 2 | | | | | | |
| 3 | Student | Score | | Grades | Bins | |
| 4 | Sal | 87 | | F | 60 | |
| 5 | Sali | 94 | | D | 70 | |
| 6 | Sally | 72 | | C | 80 | |
| 7 | Sam | 82 | | B | 90 | |
| 8 | Sara | 88 | | A | 100 | |
| 9 | Sarah | 78 | | | | |
| 10 | Su Nee | 95 | | | | |
| 11 | Stewart | 91 | | | | |
| 12 | Stuart | 68 | | | | |
| 13 | Suri | 84 | | | | |
| 14 | | | | | | |

The values in the bin column are the upper limits of each bin, so a score of 90 would be given a grade of B, not A. Since it is common for a score of 90 to receive an A grade, we can modify the values in the bins slightly, as shown in Figure 5.8.

To create a histogram from the scores and bins shown in Figure 5.8, follow these steps:

**Figure 5.8**
Modified bin values.

| | A | B | C | D | E | F |
|---|---|---|---|---|---|---|
| 1 | Grade Histogram | | | | | |
| 2 | | | | | | |
| 3 | Student | Score | | Grades | Bins | |
| 4 | Sal | 87 | | F | 59.99 | |
| 5 | Sali | 94 | | D | 69.99 | |
| 6 | Sally | 72 | | C | 79.99 | |
| 7 | Sam | 82 | | B | 89.99 | |
| 8 | Sara | 88 | | A | 100 | |
| 9 | Sarah | 78 | | | | |
| 10 | Su Nee | 95 | | | | |
| 11 | Stewart | 91 | | | | |
| 12 | Stuart | 68 | | | | |
| 13 | Suri | 84 | | | | |
| 14 | | | | | | |

1. Open the Data Analysis dialog box (shown in Figure 5.9) with Ribbon options: **Data tab → Analysis group → Data Analysis button**. (Excel 2003: Use menu options **Tools → Data Analysis**.)
2. Select **Histogram** from the list of data analysis tools (illustrated in Figure 5.9).
3. Click **OK** to close the Data Analysis dialog box and open the Histogram dialog box (shown in Figure 5.10).
4. Select the score values (cells B4:B13) as the **Input Range** (shown in Figure 5.10).
5. Select the bin values (cells E4:E8) as the **Bin Range** (shown in Figure 5.10).
6. Select where you want the output to be located. In this example we have chosen to place the output below the data by entering A16 as the top-left cell of the **Output Range** (shown in Figure 5.10).
7. Request the histogram chart by checking the **Chart Output** box (shown in Figure 5.10).
8. Click **OK** to close the Histogram dialog box and create the histogram chart.

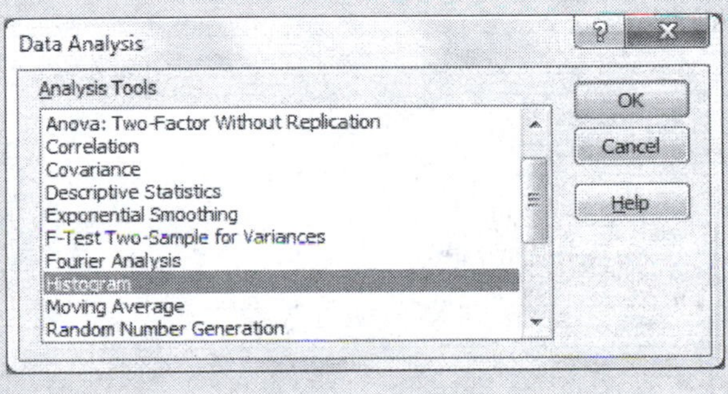

**Figure 5.9**
The Data Analysis dialog box with the Histogram tool selected.

Data Analysis

Analysis Tools

Anova: Two-Factor Without Replication
Correlation
Covariance
Descriptive Statistics
Exponential Smoothing
F-Test Two-Sample for Variances
Fourier Analysis
Histogram
Moving Average
Random Number Generation

OK
Cancel
Help

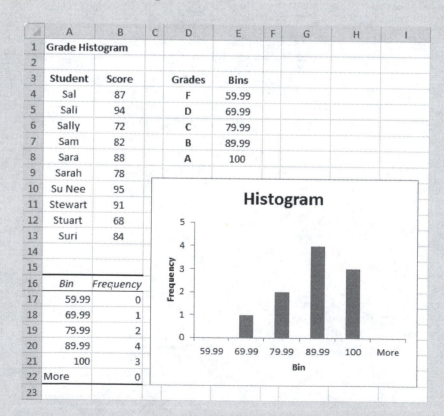

**Figure 5.10**
The Histogram dialog box, with required fields filled in.

The result is shown in Figure 5.11.

**Figure 5.11**
The resulting histogram
(chart enlarged slightly).

If you want to replace the score values on the *x*-axis with grade letters, just change the horizontal axis labels as follows:

1. Click the chart to select it. This causes the Ribbon to display the Chart Tools tabs.
2. Click the **Select Data** button on the Chart Tools: Design tab (Ribbon options: **Chart Tools → Design tab → Data group → Select Data button**. This opens the Select Data Source dialog box shown in Figure 5.12.
3. Click the **Edit** button in the Horizontal (Category) Axis Labels area (indicated in Figure 5.12).
4. Select the cells containing the letter grades (cells D4:D8) as shown in Figure 5.13.
5. Click **OK** to close the Axis Labels dialog box (Figure 5.13).
6. Click **OK** to close Select Data Source dialog box (Figure 5.12).

The resulting histogram, with the *x*-axis title changed from "bin" to "Grade," is shown in Figure 5.14.

**Figure 5.12**
The Select Data Source dialog box.

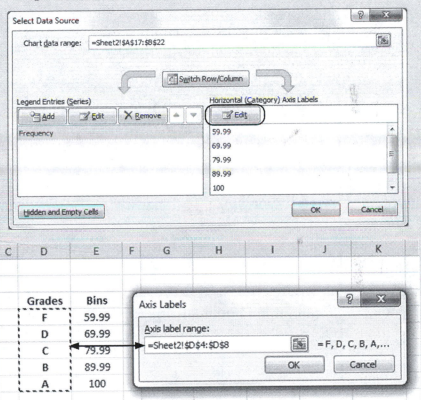

**Figure 5.13**
Select the new *x*-axis labels.

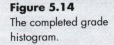

**Figure 5.14**
The completed grade histogram.

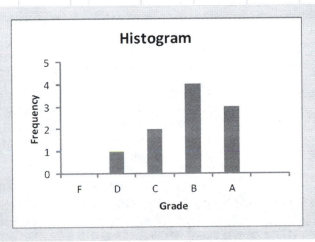

APPLICATION

## HISTOGRAMS FOR QUALITY ASSURANCE

An example of the use of a histogram is to visualize the distribution of the tolerance errors of a test set of machine parts coming off an assembly line. By a glance at the histogram, the QA department can tell whether the tolerance errors are widely or narrowly distributed. In addition, you can quickly see whether the tolerance errors are normally distributed or skewed.

The data in Figure 5.15 are the results of a test batch from an assembly line. The QA team collected 25 sample parts and measured their tolerances. The team noted that the test data typically range from about −5 to +5 thousandths of an inch from the correct size. The team created nine bins for the data to fall into.

**Figure 5.15**

Tolerances of test batch machine parts.

| | A | B | C | D | E | F |
|---|---|---|---|---|---|---|
| 1 | **Machine Parts Assembly Line 14** | | | | | |
| 2 | | | | | | |
| 3 | Tolerance | Bins | | Test Batch # | 3201 | |
| 4 | 0.34 | -3.5 | | Date: | 2/1/2008 | |
| 5 | 1.03 | -2.5 | | Units: | inches x 10³ | |
| 6 | -1.26 | -1.5 | | | | |
| 7 | 3.13 | -0.5 | | | | |
| 8 | -0.1 | 0.5 | | | | |
| 9 | 0.02 | 1.5 | | | | |
| 10 | -0.01 | 2.5 | | | | |
| 11 | 2.12 | 3.5 | | | | |
| 12 | -1.4 | 4.5 | | | | |
| 13 | 1.24 | | | | | |
| 14 | 2.29 | | | | | |
| 15 | -0.71 | | | | | |
| 16 | -1.38 | | | | | |
| 17 | -1.13 | | | | | |
| 18 | -1.34 | | | | | |
| 19 | 0.03 | | | | | |
| 20 | -0.03 | | | | | |
| 21 | -0.56 | | | | | |
| 22 | -0.04 | | | | | |
| 23 | -2.55 | | | | | |
| 24 | -0.01 | | | | | |
| 25 | 0.56 | | | | | |
| 26 | 0.78 | | | | | |
| 27 | 0.99 | | | | | |
| 28 | 1.12 | | | | | |
| 29 | | | | | | |

To create a histogram of the data in Figure 5.15, perform the following steps:

1. Open the Data Analysis dialog box with Ribbon options: **Data tab → Analysis group → Data Analysis button**. (Excel 2003: Use menu options **Tools → Data Analysis**).
2. Select **Histogram** from the list of data analysis tools.
3. Click **OK** to close the Data Analysis dialog box and open the Histogram dialog box.
4. Select the tolerance values (cells A3:A28) as the **Input Range** (shown in Figure 5.16).

5.  Select the bin values (cells B3:B12) as the **Bin Range** (shown in Figure 5.16).
6.  Notice that the cell ranges used as the input Range and Bin Range included the column headings ("Tolerance" and "Bins"). Check the box labeled **Labels** (shown in Figure 5.16) to let Excel know that labels were included in the specified ranges.
7.  Select where you want the output to be located. In this example we have chosen to send the output to a **New Worksheet Ply** (shown in Figure 5.16).
8.  Request the histogram chart by checking the **Chart Output** box (shown in Figure 5.16).
9.  Click **OK** to close the Histogram dialog box and create the histogram chart.

The result is shown in Figure 5.17. You could use the Chart Tools on the Ribbon to change the appearance of the histogram, if needed.

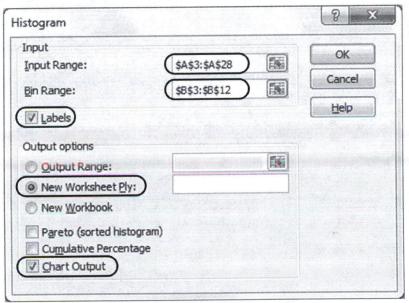

**Figure 5.16**
The Histogram dialog box filled in for the machine part tolerances histogram.

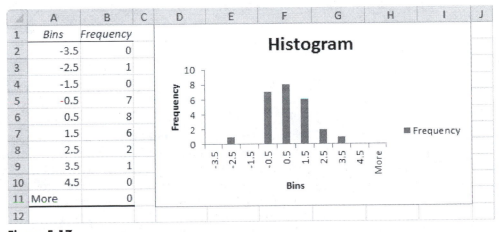

**Figure 5.17**
Histogram of tolerances of tested machine parts.

Several other features are available on the Histogram dialog box (see Figure 5.16). These include sorting the histogram (*Pareto*) and displaying a cumulative percentage (*Cumulative Percentage*).

The **Help** button on the Histogram dialog box displays detailed information about the input parameters and options for the Histogram tool.

## 5.4 PROVIDING DESCRIPTIVE STATISTICS

Excel provides a data analysis tool for computing common ***descriptive statistics***, which are values calculated from a data set and used to describe some basic characteristics of the data set. You could calculate each of these values by using individual Excel built-in functions from the Insert Function dialog box, but the Descriptive Statistics tool conveniently aggregates the most common computations for you in a table.

As an example, we will go through the steps to compute the descriptive statistics of the machine parts test data from Figure 5.15:

1. Open the Data Analysis dialog box with Ribbon options: **Data tab → Analysis group → Data Analysis button**. (Excel 2003: Use menu options Tools → Data Analysis.)
2. Select **Descriptive Statistics** from the list of data analysis tools.
3. Click **OK** to close the Data Analysis dialog box and open the Descriptive Statistics dialog box, shown in Figure 5.18.
4. Select the tolerance values (cells A3:A28) as the **Input Range** (shown in Figure 5.18).
5. Since the Input Range included a column heading, check the **Labels in First Row** box (shown in Figure 5.18).
6. Choose a location for the output table. In this example, we have opted to send the output to a **New Worksheet Ply**.
7. Check the **Summary statistics** box (shown in Figure 5.18) to tell Excel to calculate the descriptive statistics.
8. Click **OK** to close the Descriptive Statistics dialog box and calculate the statistic values.

The results are shown in Figure 5.19.

You are not expected to understand all of the terms in Figure 5.19. However, you can probably recognize some of them. Each statistic is described briefly in Table 5.1.

From these results, the QA team can see that the measured tolerances ranged from −2.55 to +3.13 thousandths of an inch, with a mean tolerance close to zero (0.1252 thousandths of an inch). An important result for the team is the variability of the test data. The standard deviation (1.305 thousandths of an inch) is a measure of the data's variability. The standard deviation of the test data set, along with the size of the data set, can be used to make predictions about how many of the items on the assembly line will eventually be rejected.

**Figure 5.18**
Filling out the Descriptive
Statistics dialog box.

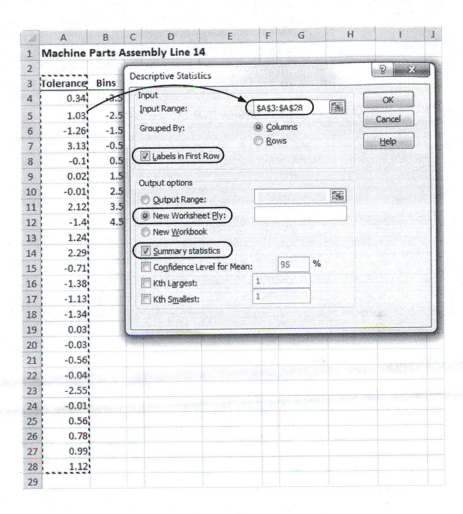

|    | A | B | C |
|----|---|---|---|
| 1  | **Machine Parts Assembly Line 14** | | |
| 2  | | | |
| 3  | Tolerance | Bins | |
| 4  | 0.34 | -3.5 | |
| 5  | 1.03 | -2.5 | |
| 6  | -1.26 | -1.5 | |
| 7  | 3.13 | -0.5 | |
| 8  | -0.1 | 0.5 | |
| 9  | 0.02 | 1.5 | |
| 10 | -0.01 | 2.5 | |
| 11 | 2.12 | 3.5 | |
| 12 | -1.4 | 4.5 | |
| 13 | 1.24 | | |
| 14 | 2.29 | | |
| 15 | -0.71 | | |
| 16 | -1.38 | | |
| 17 | -1.13 | | |
| 18 | -1.34 | | |
| 19 | 0.03 | | |
| 20 | -0.03 | | |
| 21 | -0.56 | | |
| 22 | -0.04 | | |
| 23 | -2.55 | | |
| 24 | -0.01 | | |
| 25 | 0.56 | | |
| 26 | 0.78 | | |
| 27 | 0.99 | | |
| 28 | 1.12 | | |
| 29 | | | |

Descriptive Statistics

Input
Input Range: $A$3:$A$28
Grouped By: ● Columns ○ Rows
☑ Labels in First Row

Output options
○ Output Range:
● New Worksheet Ply:
○ New Workbook
☑ Summary statistics
☐ Confidence Level for Mean: 95 %
☐ Kth Largest: 1
☐ Kth Smallest: 1

OK
Cancel
Help

**Figure 5.19**
Descriptive statistics for the
tested machine parts.

|    | A | B | C |
|----|---|---|---|
| 1  | *Tolerance* | | |
| 2  | | | |
| 3  | Mean | 0.1252 | |
| 4  | Standard Error | 0.261001 | |
| 5  | Median | -0.01 | |
| 6  | Mode | -0.01 | |
| 7  | Standard Deviation | 1.305004 | |
| 8  | Sample Variance | 1.703034 | |
| 9  | Kurtosis | 0.212016 | |
| 10 | Skewness | 0.296037 | |
| 11 | Range | 5.68 | |
| 12 | Minimum | -2.55 | |
| 13 | Maximum | 3.13 | |
| 14 | Sum | 3.13 | |
| 15 | Count | 25 | |
| 16 | | | |

**Table 5.1 The Descriptive Statistics**

| Statistic | Value* | Description |
|---|---|---|
| Mean | 0.1252 | The data set shows scatter around a central point, which is called the *mean*. |
| Standard Error | 0.261001 | The *standard error* is used to estimate how accurately the *sample mean* (0.1252 in this example) represents the *population mean*. |
| Median | −0.01 | The *median* is the middle value of the sorted data set. |
| Mode | −0.01 | The *mode* is the most commonly occurring value in the data set. |
| Standard Deviation | 1.305004 | The *standard deviation* is a measure of the data's variability. |
| Sample Variance | 1.703034 | The *sample variance* is the square of the *sample standard deviation*. |
| Kurtosis | 0.212016 | The *kurtosis* provides information about the "peakedness" of the data distribution compared with a normal distribution. Large positive values of kurtosis indicate that the data distribution is "tall and skinny" compared with the normal distribution. |
| Skewness | 0.296037 | The *skewness* provides information about the extent of asymmetry of the distribution about its *mean*. |
| Range | 5.68 | The *range* is the difference between the maximum and minimum values in the data set. |
| Minimum | −2.55 | *Minimum* indicates the smallest value in the data set. |
| Maximum | 3.13 | *Maximum* indicates the largest value in the data set. |
| Sum | 3.13 | The *sum* is the result of adding all values in the data set. |
| Count | 25 | The *count* is the number of values in the data set. |

*Values are from Figure 5.19.

## COMPARING TWO ASSEMBLY LINES

APPLICATION

In Figure 5.15, test data for a single assembly line were presented. In this example, we will compare the results from two assembly lines. The collected data are presented in Figure 5.20.

In Figure 5.20, new bins have been assigned to cover the range of values from assembly line #17. Using this data, a histogram for each line can be prepared. The results are shown in Figure 5.21.

The histograms in Figure 5.21 clearly show that Line 17 is producing parts with a much wider range of tolerances. Therefore, there is a good chance that Line 17 is producing more rejected parts than Line 14. The engineer responsible for Line 17 will probably start looking for reasons why the distribution of tolerances is so wide, so that the equipment can be repaired.

**Figure 5.20**

Test data from two assembly lines.

| | A | B | C | D | E | F | G |
|---|---|---|---|---|---|---|---|
| 1 | **Machine Parts Tolerances** | | | | | | |
| 2 | | | | | | | |
| 3 | Line 14 | Line 17 | Bins | | Test Batch # 3201 | | |
| 4 | 0.34 | 1.14 | -10 | | Date: 2/1/2008 | | |
| 5 | 1.03 | 0.2 | -8 | | Units: inches x $10^3$ | | |
| 6 | -1.26 | -0.31 | -6 | | | | |
| 7 | 3.13 | 4.84 | -4 | | | | |
| 8 | -0.1 | -1.12 | -2 | | | | |
| 9 | 0.02 | 0.12 | 0 | | | | |
| 10 | -0.01 | 9.35 | 2 | | | | |
| 11 | 2.12 | -5.8 | 4 | | | | |
| 12 | -1.4 | -0.31 | 6 | | | | |
| 13 | 1.24 | 6.02 | 8 | | | | |
| 14 | 2.29 | 0.21 | 10 | | | | |
| 15 | -0.71 | -9.72 | | | | | |
| 16 | -1.38 | 1.88 | | | | | |
| 17 | -1.13 | 4.72 | | | | | |
| 18 | -1.34 | -1.01 | | | | | |
| 19 | 0.03 | -5.72 | | | | | |
| 20 | -0.03 | 1.86 | | | | | |
| 21 | -0.56 | -7.02 | | | | | |
| 22 | -0.04 | 4.49 | | | | | |
| 23 | -2.55 | -5.27 | | | | | |
| 24 | -0.01 | -2.99 | | | | | |
| 25 | 0.56 | -5.71 | | | | | |
| 26 | 0.78 | 2.36 | | | | | |
| 27 | 0.99 | -5.76 | | | | | |
| 28 | 1.12 | -1.14 | | | | | |
| 29 | | | | | | | |

**Figure 5.21**

Histograms of the data from each assembly line.

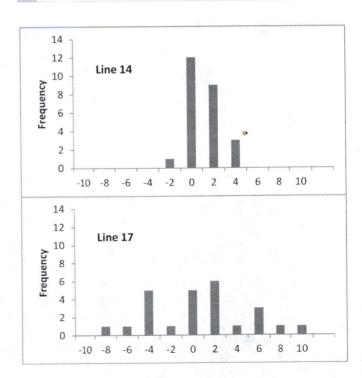

To quantify the differences between the two data sets, descriptive statistics for the two sets can be calculated. To compare the values side by side, simply include both data sets in the Input Range on the Descriptive Statistics dialog box. This is illustrated in Figure 5.22. The calculated results are shown in Figure 5.23.

**Figure 5.22**

The Descriptive Statistics dialog box with two data columns in the Input Range.

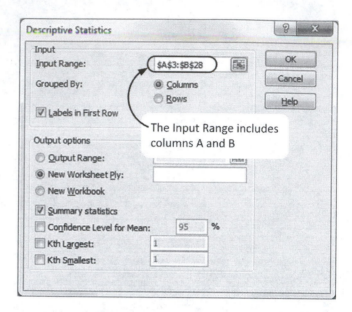

**Figure 5.23**

Comparing the descriptive statistics for the two assembly lines.

| | A | B | C | D | E |
|---|---|---|---|---|---|
| 1 | *Line 14* | | *Line 17* | | |
| 2 | | | | | |
| 3 | Mean | 0.1252 | Mean | -0.5876 | |
| 4 | Standard Error | 0.261001 | Standard Error | 0.92307 | |
| 5 | Median | -0.01 | Median | -0.31 | |
| 6 | Mode | -0.01 | Mode | -0.31 | |
| 7 | Standard Deviation | 1.305004 | Standard Deviation | 4.615351 | |
| 8 | Sample Variance | 1.703034 | Sample Variance | 21.30146 | |
| 9 | Kurtosis | 0.212016 | Kurtosis | -0.32901 | |
| 10 | Skewness | 0.296037 | Skewness | 0.035812 | |
| 11 | Range | 5.68 | Range | 19.07 | |
| 12 | Minimum | -2.55 | Minimum | -9.72 | |
| 13 | Maximum | 3.13 | Maximum | 9.35 | |
| 14 | Sum | 3.13 | Sum | -14.69 | |
| 15 | Count | 25 | Count | 25 | |
| 16 | | | | | |

In Figure 5.23 we see that the mean tolerances for the two assembly lines are different, and that the standard deviation of the values from Line 17 is much larger than the standard deviation from Line 14. Since the standard deviation is a measure of the data's variability, this result confirms (and quantifies) what was evident in the histograms: The distribution of tolerances in Line 17 is much wider.

## 5.5 COMPUTING A CORRELATION

The *correlation* of two data sets is a measure of how well the two ranges of data move together linearly. If large values of one set are associated with large values of the other, then a positive correlation exists. If small values of one set are associated with large values of the other (and vice versa), then a negative correlation exists.

In the left chart in Figure 5.24, the two curves increase together; these data sets are strongly and positively correlated. On the other hand, the data sets shown in the right chart show a strong negative correlation; they move together, but not in the same direction.

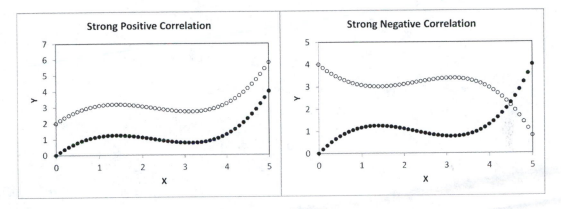

**Figure 5.24**

Examples of correlated data sets. Left panel: positive correlation, right panel: negative correlation.

If the correlation is near zero, then the values in both sets are not linearly related (Figure 5.25).

**Figure 5.25**
Weakly correlated data.

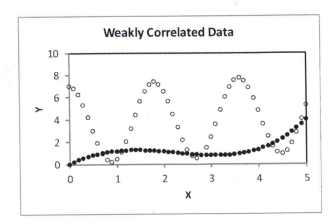

Figure 5.26 shows the midterm grades and overall grade point averages (GPAs) for 20 students. We are interested in knowing whether there is a positive correlation between the midterm exam grades and the students' GPAs.

**Figure 5.26**
Midterm scores and GPAs
for 20 students.

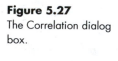

| Student # | Midterm Score | GPA |
|---|---|---|
| 1 | 45 | 1.7 |
| 2 | 89 | 3.7 |
| 3 | 90 | 3.9 |
| 4 | 67 | 3.7 |
| 5 | 88 | 2.4 |
| 6 | 93 | 3.4 |
| 7 | 32 | 1.8 |
| 8 | 85 | 3.1 |
| 9 | 68 | 2.4 |
| 10 | 52 | 2.6 |
| 11 | 77 | 3.5 |
| 12 | 96 | 3.9 |
| 13 | 54 | 2.1 |
| 14 | 78 | 2.8 |
| 15 | 83 | 2.4 |
| 16 | 89 | 3.1 |
| 17 | 79 | 2.9 |
| 18 | 83 | 2.9 |
| 19 | 72 | 3.1 |
| 20 | 91 | 3.6 |

Once the data are available in an Excel worksheet, we can calculate the correlation between the students' midterm grades and GPAs with these steps:

1. Open the Data Analysis dialog box with Ribbon options: **Data tab → Analysis group → Data Analysis button**. (Excel 2003: Use menu options Tools → Data Analysis.)
2. Select **Correlation** from the list of data analysis tools.
3. Click **OK** to close the Data Analysis dialog box and open the Correlation dialog box, shown in Figure 5.27.
4. Use the student midterm scores and GPAs (cells B1:C21, including the column headings) as the **Input Range** for the calculation (shown in Figure 5.27).
5. Since the Input Range included column headings, check the **Labels in First Row** box (shown in Figure 5.27).

**Figure 5.27**
The Correlation dialog
box.

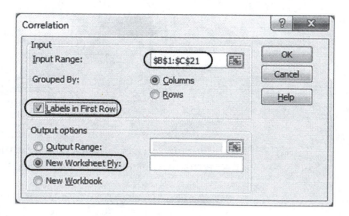

6. Choose a location for the output table. In this example, we have opted to send the output to a **New Worksheet Ply**.

7. Click **OK** to close the Correlation dialog box and calculate the *correlation coefficient, r*.

The result is shown in Figure 5.28. Note that the correlation coefficient $r = 0.7354$ indicates that there is a positive correlation between midterm grades and GPAs for this group of students.

**Figure 5.28**
The correlation between midterm scores and GPAs.

| | A | B | C | D |
|---|---|---|---|---|
| 1 | | *Midterm Score* | GPA | |
| 2 | Midterm Score | 1 | | |
| 3 | GPA | 0.7354 | 1 | |
| 4 | | | | |

## 5.6 PERFORMING A LINEAR REGRESSION

The *correlation coefficient* is a measure of whether a linear relationship exists between two sets of data. However, the correlation coefficient does not tell us what the relationship is, merely that it exists. A *linear regression analysis* is an attempt to find the relationship among variables and express the relationship as a linear equation.

The Analysis ToolPak performs linear regression analysis by using the least squares method to fit a curve through a set of observations. In this section, we will show you how to perform a *regression* that analyzes how a single dependent variable is affected by a single independent variable.

If the regression is a good fit to the data, then it allows us to make predictions of future performance. For example, we might make a prediction about a student's first-year college GPA on the basis of the student's high school GPA.

To do so, we would follow these steps:

1. Collect GPA data from a representative sample of college students. The data would include their high school GPA and their first-year college GPA.

2. Then, we would perform a linear regression, using the high school and college GPA data. The result of the regression would be a linear equation relating college GPA to high school GPA.

3. Next, we would use the regression equation with the GPAs of a group of recent high school graduates to predict how they would do in their first year of college.

If more than one independent variable is considered, then the analysis technique is called *multiple regression*. For example, the student's SAT score and IQ might both be tested as predictors, in addition to the high school GPA.

Since we know that there is a positive relationship between overall GPA and midterm scores in the data in Figure 5.26, let's perform a regression analysis on the same data. We would like to know whether we can predict a student's midterm grade, given the student's GPA. We are making some strong assumptions about our collected data. These are assumptions about our sampling technique and the underlying distribution of the student GPAs. We will ignore these ramifications here, but you will study them if you take a statistics course.

To perform a regression analysis, follow these steps:

1. Open the Data Analysis dialog box with Ribbon options: **Data tab → Analysis group → Data Analysis button**. (Excel 2003: Use menu options Tools → Data Analysis.)
2. Select **Regression** from the list of data analysis tools.
3. Click **OK** to close the Data Analysis dialog box and open the Regression dialog box, shown in Figure 5.29.
4. Use the student midterm scores (cells B1:B21, including the column heading) as the **Input Y Range** for the calculation (shown in Figure 5.29).
5. Use the student GPAs (cells C1:C21, including the column heading) as the **Input X Range** for the calculation (shown in Figure 5.29).

**Note:** In regression analysis, *x* is used to predict *y*. Since we are trying to predict midterm score from GPA, the midterm scores are the **Input Y Range** and the GPAs are the **Input X Range**.

1. Since the input ranges included column headings, check the **Labels** box (shown in Figure 5.29).
2. Choose a location for the output table. In this example, we have opted to send the output to a **New Worksheet Ply**.
3. Click **OK** to close the Regression dialog box and perform the regression analysis.

The results are shown in Figure 5.30.

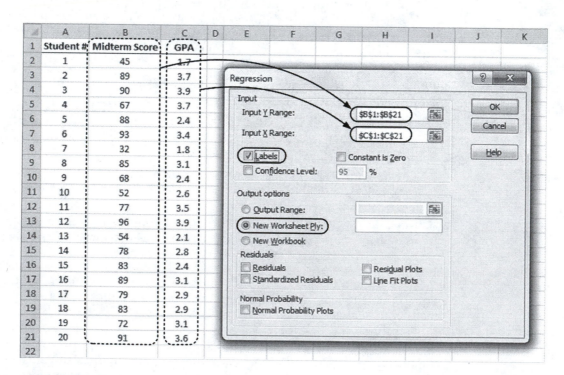

**Figure 5.29**
The Regression dialog box being filled in.

**Figure 5.30**
Regression analysis results.

| | A | B | C | D | E | F | G |
|---|---|---|---|---|---|---|---|
| 1 | SUMMARY OUTPUT | | | | | | |
| 2 | | | | | | | |
| 3 | *Regression Statistics* | | | | | | |
| 4 | Multiple R | 0.735 | | | | | |
| 5 | R Square | 0.541 | | | | | |
| 6 | Adjusted R Square | 0.515 | | | | | |
| 7 | Standard Error | 12.266 | | | | | |
| 8 | Observations | 20 | | | | | |
| 9 | | | | | | | |
| 10 | ANOVA | | | | | | |
| 11 | | *df* | *SS* | *MS* | *F* | *Significance F* | |
| 12 | Regression | 1 | 3190.55 | 3190.55 | 21.20 | 0.0002 | |
| 13 | Residual | 18 | 2708.40 | 150.47 | | | |
| 14 | Total | 19 | 5898.95 | | | | |
| 15 | | | | | | | |
| 16 | | *Coefficients* | *Standard Error* | *t Stat* | *P-value* | | |
| 17 | Intercept | 18.696 | 12.648 | 1.478 | 0.157 | | |
| 18 | GPA | 19.272 | 4.185 | 4.605 | 0.0002 | | |
| 19 | | | | | | | |

Figure 5.30 presents a lot more information than you probably care to know. One of the important results is the significance of the $F$ statistic (Significance $F$ = 0.0002). This means that there is a very low probability that the results are from chance.

The regression coefficients (Intercept and GPA) are listed in the *Coefficients* column. We use these to build a predictive equation:

$$\text{Predicted Midterm Grade} = 19.272 \times \text{GPA} + 18.696$$

The $R^2$ value ($R^2 = 0.541$) is a pretty low value, considering that $R^2 = 1.000$ indicates that the regression line is a perfect fit to the data. Also, notice that the *standard error of the intercept* (12.648) is almost as great as the value of the coefficient (18.696). Ideally, you would like to see standard errors much smaller than the coefficients. Finally, look at the *P-values* associated with the coefficients in Figure 5.30. The *P*-values represent the probability that you could get the calculated result by random chance. Ideally, you would want very small *P*-values associated with your coefficients, such as the $P = 0.0002$ on the GPA. This indicates that the probability that the GPA coefficient could be obtained by random chance is only 0.02%. However, the probability that the intercept coefficient could be obtained by random chance is 15.7%. That's pretty high, and it suggests that there is a lot of uncertainty associated with the intercept coefficient in the regression.

When a coefficient has a high *P*-value, you have to wonder if the coefficient needs to be included in the model at all. We can rework the regression analysis without an intercept (actually, setting the intercept to zero) to see if it makes a difference in the results. To do so, follow the same check, but check the **Constant is Zero** box on the Regression dialog box, as shown in Figure 5.31. The results of performing the regression with the intercept forced to be equal to zero are shown in Figure 5.32.

**Figure 5.31**
Forcing the intercept equal to zero on the Regression dialog box.

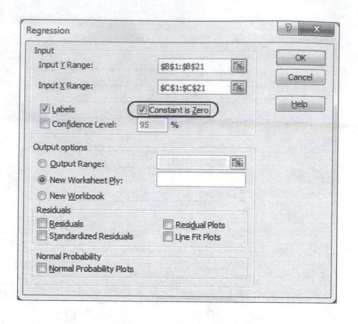

**Figure 5.32**
Regression results with intercept set equal to zero.

|   | A | B | C | D | E | F | G |
|---|---|---|---|---|---|---|---|
| 1 | SUMMARY OUTPUT | | | | | | |
| 2 | | | | | | | |
| 3 | *Regression Statistics* | | | | | | |
| 4 | Multiple R | 0.987 | | | | | |
| 5 | R Square | 0.975 | | | | | |
| 6 | Adjusted R Square | 0.922 | | | | | |
| 7 | Standard Error | 12.643 | | | | | |
| 8 | Observations | 20 | | | | | |
| 9 | | | | | | | |
| 10 | ANOVA | | | | | | |
| 11 | | *df* | *SS* | *MS* | *F* | *Significance F* | |
| 12 | Regression | 1 | 117017.79 | 117017.79 | 732.03 | 4.952E-16 | |
| 13 | Residual | 19 | 3037.21 | 159.85 | | | |
| 14 | Total | 20 | 120055 | | | | |
| 15 | | | | | | | |
| 16 | | *Coefficients* | *Standard Error* | *t Stat* | *P-value* | | |
| 17 | Intercept | 0 | #N/A | #N/A | #N/A | | |
| 18 | GPA | 25.312 | 0.936 | 27.056 | 1.23E-16 | | |
| 19 | | | | | | | |

Notice that the $R^2$ value is much closer to 1.0, and the Significance $F$ and $P$-value for the GPA coefficient are minuscule. This appears to be a better result. The predictive equation is now

$$\text{Predicted Midterm Grade} = 25.312 \times \text{GPA}$$

Figure 5.33 shows both regression equations superimposed on a graph of the original data.

**Figure 5.33**
The regression equations
with the original data.

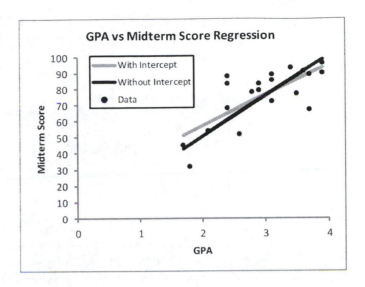

## 5.7 TREND ANALYSIS

***Trend analysis*** is the science of forecasting or predicting future elements of a data series on the basis of historical data. Trend analysis is used in many areas, such as financial forecasting, epidemiology, capacity planning, and criminology. Excel has the ability to calculate linear and ***exponential growth*** trends for data series. Excel can also calculate and display various trendlines for charts.

A ***linear trend*** consists of fitting known *x* and *y* values to a linear equation. Unknown *y* values may then be predicted by extending the *x* values and using the equation to calculate new *y* values. This is called *linear extension*. An exponential trend can be created in a similar manner, except that the data are fit to an exponential equation. The graph of a trend is called a ***trendline***.

### 5.7.1 Trend Analysis with Data Series

A trend analysis can either extend or replace a series of data elements. The simplest method for extending a data series with a linear regression is to drag the Fill Handle past the end of the data series. For example, Figure 5.34 shows the number of occurrences of a hypothetical disease for the years 2007 to 2010.

**Figure 5.34**
Disease data.

| | A | B | C | D | E | F | G | H | I | J |
|---|---|---|---|---|---|---|---|---|---|---|
| 1 | Occurrence of Disease X (in thousands) | | | | | | | | | |
| 2 | | | | | | | | | | |
| 3 | | Known Values | | | | Predicted Values | | | | |
| 4 | 2007 | 2008 | 2009 | 2010 | 2011 | 2012 | 2013 | 2014 | 2015 | |
| 5 | 1.1 | 1.9 | 3.0 | 3.8 | | | | | | |
| 6 | | | | | | | | | | |

By assuming a linear rate of increase in the disease, the number of occurrences can be estimated for years 2011–2015. To calculate the *linear trend*:

1. Select the known data (A5:D5).
2. Then, drag the Fill Handle to the right so that the fill box covers cells (E5:I5).

When you release the mouse, cells (E5:I5) will contain the data elements predicted by a linear regression of the original data. This result is shown in Figure 5.35.

**Figure 5.35**

Disease data and projected disease occurrences.

| A | B | C | D | E | F | G | H | I | J |
|---|---|---|---|---|---|---|---|---|---|
| 1 | Occurrence of Disease X (in thousands) | | | | | | | | |
| 2 | | | | | | | | | |
| 3 | | Known Values | | | | Predicted Values | | | |
| 4 | 2007 | 2008 | 2009 | 2010 | 2011 | 2012 | 2013 | 2014 | 2015 |
| 5 | 1.1 | 1.9 | 3.0 | 3.8 | 4.8 | 5.7 | 6.6 | 7.5 | 8.4 |
| 6 | | | | | | | | | |

Notice that the date values in row 4 were never used when calculating the linear trend and filling cells E5:I5 by using the Fill Handle. Excel assumes that the known values (A5:D5) are separated by equal intervals (one year in this example), and that the new values will also have the same interval. Thus, you can predict what will happen in 2011, 2012, and 2013, but you cannot use the Fill Handle to predict what will happen in 2011, 2020, and 2050 (unless you calculate all intervening years as well).

The *Fill Series command* can be used for a somewhat more sophisticated trend analysis. To extend or replace a data series by using the Fill Series command, first select the region of data over which the analysis is to occur. This includes the original data and the new cells that are to hold the predicted data. Let's try some of the Fill Series functions, using the sample data in Figure 5.31:

1. Create a worksheet with the titles and headers shown in Figure 5.36.
2. Type the numbers 1.1, 1.9, 3.0, and 3.8 into cells B6:E6.
3. Copy the values to rows 8, 10, 12, and 14—the same known values will be used for each trend analysis.

**Figure 5.36**

Preparing to use the Fill Series command.

| | A | B | C | D | E | F | G | H | I | J | K |
|---|---|---|---|---|---|---|---|---|---|---|---|
| 1 | Using the Fill Series Command | | | | | | | | | | |
| 2 | | | | Occurrence of Disease X (in thousands) | | | | | | | |
| 3 | | | Known Values | | | | Predicted Values | | | | |
| 4 | Type of Series | 2007 | 2008 | 2009 | 2010 | 2011 | 2012 | 2013 | 2014 | 2015 | |
| 5 | | | | | | | | | | | |
| 6 | Original Known Values | 1.1 | 1.9 | 3.0 | 3.8 | | | | | | |
| 7 | | | | | | | | | | | |
| 8 | Linear Extension | 1.1 | 1.9 | 3.0 | 3.8 | | | | | | |
| 9 | | | | | | | | | | | |
| 10 | Linear Replacement | 1.1 | 2.0 | 2.9 | 3.9 | | | | | | |
| 11 | | | | | | | | | | | |
| 12 | Exponential Replacement | 1.2 | 1.8 | 2.7 | 4.1 | | | | | | |
| 13 | | | | | | | | | | | |
| 14 | *TREND* Function | 1.1 | 1.9 | 3.0 | 3.8 | | | | | | |
| 15 | | | | | | | | | | | |
| 16 | | | | | | | | | | | |

### 5.7.2 Trend Analysis: Linear Extension

To extend the known values with a linear regression and leave the original values unchanged, follow these steps:

1. Select cells B8:J8, the known data points and the unknown (to be calculated) points.
2. Open the Series dialog box (shown in Figure 5.37) using Ribbon options: **Home tab → Editing group → Fill drop-down menu → Series . . . button.** (Excel 2003: Edit → Fill → Series.)

   Note that a step value (the amount by which a series is increased or decreased) has been calculated by Excel. The step value can be modified manually to set the increment value for *x* in the linear equation

   $$y = mx + b.$$

A stop value (the value at which the series is to end) may be entered if you want to set an upper limit to the trend.

3. Check the box labeled "Rows" in the **Series in** group.
4. Check the box labeled **AutoFill**.
5. Clear all other checkboxes.

The results are depicted in Row 8 in Figure 5.38 (labeled Linear Extension). Note that these results are identical to the results obtained by dragging the Fill Handle in Figure 5.35.

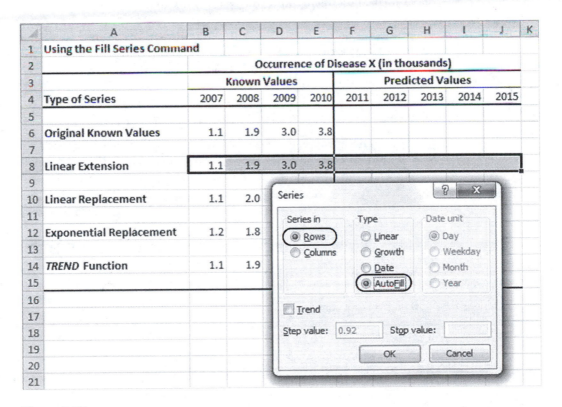

**Figure 5.37**
The Series dialog box in use.

**Figure 5.38**
After performing the linear extension trend analysis.

| | 2007 | 2008 | 2009 | 2010 | 2011 | 2012 | 2013 | 2014 | 2015 |
|---|---|---|---|---|---|---|---|---|---|
| **Using the Fill Series Command** | | | | | | | | | |
| **Occurrence of Disease X (in thousands)** | | | | | | | | | |
| | **Known Values** | | | | **Predicted Values** | | | | |
| **Type of Series** | 2007 | 2008 | 2009 | 2010 | 2011 | 2012 | 2013 | 2014 | 2015 |
| **Original Known Values** | 1.1 | 1.9 | 3.0 | 3.8 | | | | | |
| **Linear Extension** | 1.1 | 1.9 | 3.0 | 3.8 | 4.8 | 5.7 | 6.6 | 7.5 | 8.4 |
| **Linear Replacement** | 1.1 | 2.0 | 2.9 | 3.9 | | | | | |
| **Exponential Replacement** | 1.2 | 1.8 | 2.7 | 4.1 | | | | | |
| *TREND* **Function** | 1.1 | 1.9 | 3.0 | 3.8 | | | | | |

## 5.7.3 Trend Analysis: Linear Replacement

Follow these steps to calculate a linear trend and replace the original data values with best fit data:

1. Select cells B10:J10, the known data points and the unknown (to be calculated) points.
2. Open the Series dialog box using Ribbon options: **Home tab → Editing group → Fill drop-down menu → Series . . . button**. (Excel 2003: Edit → Fill → Series.)
3. Check the box labeled "Rows" in the **Series in** group.
4. Check the box labeled **Linear**.
5. Check the box labeled **Trend**—this causes the original data points to be replaced by values calculated using the linear trend equation.

The results are depicted in Row 10 in Figure 5.39 (labeled Linear Replacement). Note that the predicted values are the same as in the previous example, but some of the known values have been modified.

**Figure 5.39**
After linear replacement trend analysis.

| | 2007 | 2008 | 2009 | 2010 | 2011 | 2012 | 2013 | 2014 | 2015 |
|---|---|---|---|---|---|---|---|---|---|
| **Using the Fill Series Command** | | | | | | | | | |
| **Occurrence of Disease X (in thousands)** | | | | | | | | | |
| | **Known Values** | | | | **Predicted Values** | | | | |
| **Type of Series** | 2007 | 2008 | 2009 | 2010 | 2011 | 2012 | 2013 | 2014 | 2015 |
| **Original Known Values** | 1.1 | 1.9 | 3.0 | 3.8 | | | | | |
| **Linear Extension** | 1.1 | 1.9 | 3.0 | 3.8 | 4.8 | 5.7 | 6.6 | 7.5 | 8.4 |
| **Linear Replacement** | 1.1 | 2.0 | 2.9 | 3.9 | 4.8 | 5.7 | 6.6 | 7.5 | 8.5 |
| **Exponential Replacement** | 1.2 | 1.8 | 2.7 | 4.1 | | | | | |
| *TREND* **Function** | 1.1 | 1.9 | 3.0 | 3.8 | | | | | |

These values were replaced

### 5.7.4 Trend Analysis: Exponential Replacement

To create a trend, using exponential growth series, follow these steps:

1. Select cells B12:J12, the known data points and the unknown (to be calculated) points.
2. Open the Series dialog box using Ribbon options: **Home tab → Editing group → Fill drop-down menu → Series . . . button**. (Excel 2003: Edit → Fill → Series.)
3. Check the box labeled "Rows" in the **Series in** group.
4. Check the box labeled **Growth**.
5. Check the box labeled **Trend**—this causes the original data points to be replaced by values calculated using the *growth trend* equation.

The results are depicted in Row 12 in Figure 5.40 (labeled Exponential Replacement).

**Figure 5.40**

After exponential growth replacement trend analysis.

| | A | B | C | D | E | F | G | H | I | J | K |
|---|---|---|---|---|---|---|---|---|---|---|---|
| 1 | Using the Fill Series Command | | | | | | | | | | |
| 2 | | | | | Occurrence of Disease X (in thousands) | | | | | | |
| 3 | | | Known Values | | | Predicted Values | | | | | |
| 4 | Type of Series | 2007 | 2008 | 2009 | 2010 | 2011 | 2012 | 2013 | 2014 | 2015 | |
| 5 | | | | | | | | | | | |
| 6 | Original Known Values | 1.1 | 1.9 | 3.0 | 3.8 | | | | | | |
| 7 | | | | | | | | | | | |
| 8 | Linear Extension | 1.1 | 1.9 | 3.0 | 3.8 | 4.8 | 5.7 | 6.6 | 7.5 | 8.4 | |
| 9 | | | | | | | | | | | |
| 10 | Linear Replacement | 1.1 | 2.0 | 2.9 | 3.9 | 4.8 | 5.7 | 6.6 | 7.5 | 8.5 | |
| 11 | | | | | | | | | | | |
| 12 | Exponential Replacement | 1.2 | 1.8 | 2.7 | 4.1 | 6.3 | 9.5 | 14.5 | 22.0 | 33.3 | |
| 13 | | | | | | | | | | | |
| 14 | *TREND* Function | 1.1 | 1.9 | 3.0 | 3.8 | | | | | | |
| 15 | | | | | | | | | | | |
| 16 | | | | | | | | | | | |

### 5.7.5 Trend Analysis Functions

Excel provides two trend analysis functions, one for linear trend calculation and another for calculating exponential trends. These are useful if the known dependent data may change and the trendline must be recalculated frequently. The linear trend function *TREND* uses the least squares method for its calculation. The syntax for *TREND* is

TREND(Known_y's, Known_x's, New_x's, Const).

The arguments are as follows:

- *Known_y's.* The known $y$ values are the known dependent values in the linear equation $Y = mx + b$. In Figure 5.40, the known $y$ values are 1.1, 1.9, 3.0, and 3.8.
- *Known_x's.* The known $x$ values are the values of the independent variable for which the $y$ values are known. In Figure 5.40, these are the values 2007, 2008, 2009, and 2010. If the known $x$ values are omitted, then the argument is assumed to be {1,2,3,4 . . .}.
- *New_x's.* The new $x$ values are the values of the independent variable for which you want new $y$ values to be calculated. If you want the predictions for years 2011 to 2015, then select the range (F14:J14). If you want to calculate the linear trend for the whole time span (2007–2015), then select the range (B14:J14).

- ***Const.*** If the const argument is set to FALSE, then *b* is set to zero, so the equation describing the relationship between *y* and *x* becomes $Y = mx$. If the const argument is set to TRUE or omitted, then *b* is computed.

To use the ***TREND function*** without replacing the original known values, follow this procedure:

1. Select cells F14:J14, the unknown (to be calculated) points only.
2. Choose the **Insert Function** button on the Formula bar and select the Statistical category.
3. Choose the *TREND* function from the function list. The Function Arguments dialog box will appear, as shown in Figure 5.41.
4. Select or type the range (B14:E14) for the Known_y's.
5. Select or type the range (B4:E4) for the Known_x's (the years 2007–2010).
6. Select or type the range (F4:J4) as the New_x's (the years 2011–2015).
7. Don't click **OK**. Since we want the output to be an array rather than a single value, press **Ctrl + Shift + Enter**.

The results can be seen in line 14 of the worksheet shown in Figure 5.42 (labeled TREND Function).

**Figure 5.41**
The Function Arguments dialog box for the *TREND* function.

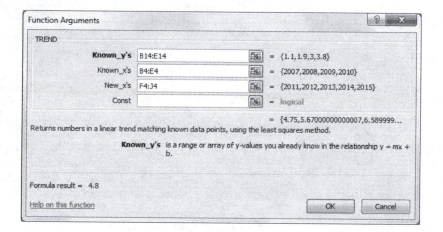

**Figure 5.42**
The result of using the *TREND* function.

| A | B | C | D | E | F | G | H | I | J | K |
|---|---|---|---|---|---|---|---|---|---|---|
| **1 Using the Fill Series Command** | | | | | | | | | | |
| **2** | | | | Occurrence of Disease X (in thousands) | | | | | | |
| **3** | | Known Values | | | | Predicted Values | | | | |
| **4 Type of Series** | 2007 | 2008 | 2009 | 2010 | 2011 | 2012 | 2013 | 2014 | 2015 | |
| **5** | | | | | | | | | | |
| **6 Original Known Values** | 1.1 | 1.9 | 3.0 | 3.8 | | | | | | |
| **7** | | | | | | | | | | |
| **8 Linear Extension** | 1.1 | 1.9 | 3.0 | 3.8 | 4.8 | 5.7 | 6.6 | 7.5 | 8.4 | |
| **9** | | | | | | | | | | |
| **10 Linear Replacement** | 1.1 | 2.0 | 2.9 | 3.9 | 4.8 | 5.7 | 6.6 | 7.5 | 8.5 | |
| **11** | | | | | | | | | | |
| **12 Exponential Replacement** | 1.2 | 1.8 | 2.7 | 4.1 | 6.3 | 9.5 | 14.5 | 22.0 | 33.3 | |
| **13** | | | | | | | | | | |
| **14 *TREND* Function** | 1.1 | 1.9 | 3.0 | 3.8 | 4.8 | 5.7 | 6.6 | 7.5 | 8.4 | |
| **15** | | | | | | | | | | |
| **16** | | | | | | | | | | |

There is a corresponding Excel function for producing an exponential growth trend, named *GROWTH*. The arguments for the **GROWTH function** are similar to the arguments for *TREND*.

## PRACTICE!

Practice using the *TREND* function to interpolate tabulated data.

Because the *TREND* function allows you to specify the $X$ values at which to predict new $Y$ values, you can use it to interpolate between values in a table. For example, steam tables contain information on various properties of water in the vapor state as functions of temperature and pressure. A section of a steam table is shown in Table 5.2.

**Table 5.2 Properties of Saturated Steam**

| Temp. °C | Enthalpy kJ/kg | Entropy kJ/kg K |
|----------|----------------|-----------------|
| 100      | 2680           | 7.4             |
| 150      | 2750           | 6.8             |
| 200      | 2790           | 6.4             |
| 250      | 2800           | 6.1             |
| 300      | 2750           | 5.7             |

Invariably, you will need to know the enthalpy at a temperature that is not included in the table, so you will have to interpolate. For example, let's find the enthalpy of saturated steam at 218°C.

To use the *TREND* function to interpolate, do the following:

1. Enter the temperature and enthalpy data for 200°C and 250°C into a worksheet, as shown in Figure 5.43.
2. Enter the desired temperature, 218°C, in the worksheet, as shown in Figure 5.43.

**Figure 5.43**
Preparing to interpolate enthalpy data.

|   | A | B | C |
|---|---|---|---|
| 1 | Interpolating Enthalpy Data | | |
| 2 | | | |
| 3 | Temp. °C | Enthalpy kJ/kg | |
| 4 | 200 | 2790 | |
| 5 | 250 | 2800 | |
| 6 | | | |
| 7 | 218 | | |
| 8 | | | |

3. Enter the TREND function in cell B7 as follows:

$$=TREND(B4:B5,A4:A5,A7)$$

The result is shown in Figure 5.44.

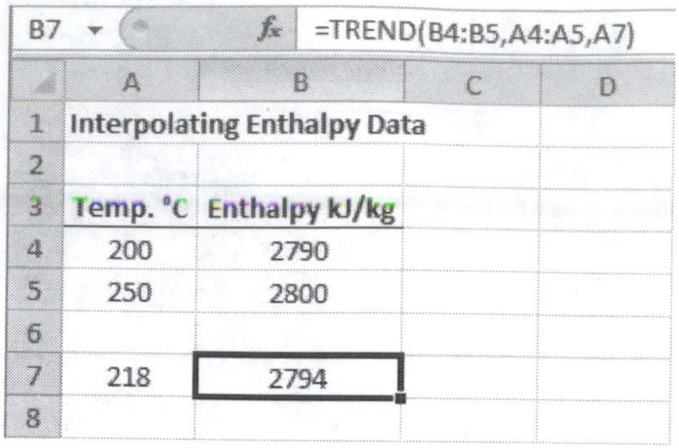

**Figure 5.44**
The result of the linear interpolation.

| B7 ▼ | | $f_x$ | =TREND(B4:B5,A4:A5,A7) | |
|---|---|---|---|---|
| ⯅ | A | B | C | D |
| 1 | Interpolating Enthalpy Data | | | |
| 2 | | | | |
| 3 | Temp. °C | Enthalpy kJ/kg | | |
| 4 | 200 | 2790 | | |
| 5 | 250 | 2800 | | |
| 6 | | | | |
| 7 | 218 | 2794 | | |
| 8 | | | | |

Practice interpolating for the following saturated steam values:

- Entropy at 218°C (Answer: 6.3 kJ/kg K)
- Enthalpy at 140°C (Answer: 2736 kJ/kg)
- Enthalpy at 350°C (Answer: 2700 kJ/kg)

### 5.7.6 Trend Analysis for Charts

Excel will calculate and display graphic representations of trends on a chart with the use of *trendlines*. Five types of regression trendlines can be added, plus a moving average trendline can be used. Each type of trendline is described in Table 5.3.

Trendlines cannot be added to all types of charts. For example, trendlines cannot be added to data series in pie charts, 3D charts, stacked charts, or doughnut charts. Trendlines can be added to Column charts, XY Scatter plots, and Line charts. If a trendline is added to a chart and the chart type is subsequently changed to one of the exempted types, then the trendline is lost.

**Table 5.3 Excel Trendline Types**

| Trendline Type | Formula |
|---|---|
| Linear | Calculates the least squares fit by using $y = mx + b$. ($m$ is the slope and $b$ is a constant.) |
| Logarithmic | Calculates least squares by using $y = c \cdot \ln(x) + b$. ($c$ and $b$ are constants.) |
| Polynomial | Calculates least squares for a line by using $y = b + c_1 x + c_2 x^2 + \cdots + c_n x^n$. <br><br>($b, c_1, c_2, \ldots, c_n$ are constants, the order can be set in the Add Trendline dialog box, and the maximum order is 6.) |
| Exponential | Calculates least squares by using $y = ce^{bx}$. ($c$ and $b$ are constants.) |
| Power | Calculates least squares by using $y = cx^b$. ($c$ and $b$ are constants.) |
| Moving Average | Calculates the series of moving averages by using <br><br>$$F_{(t+1)} = \frac{1}{N}\sum_{1}^{N}A_{t-j+1}.$$ <br><br>Keep in mind that each data point in a moving average is the average of a specified number of previous data points. $N$ is the number of prior periods to average, $A_t$ is the value at time $t$, $F_t$ is the forecasted value at time $t$. |

To create a trendline, first generate a chart of an acceptable type. As an example, we will make an XY Scatter chart by using the data in Figure 5.45. These data represent the growth of a certain type of algae in the Great Salt Lake. It has been predicted that, if left undisturbed, the algae would multiply exponentially.

**Figure 5.45**
Data Set: Growth of Algae.

|  | A | B | C |
|---|---|---|---|
| 1 | **Algae Growth Rate** | | |
| 2 | | | |
| 3 | **Day** | **Population (millions)** | |
| 4 | 1 | 1.00 | |
| 5 | 2 | 1.12 | |
| 6 | 3 | 1.92 | |
| 7 | 4 | 2.65 | |
| 8 | 5 | 4.12 | |
| 9 | 6 | 6.41 | |
| 10 | 7 | 8.66 | |
| 11 | 8 | 14.36 | |
| 12 | 9 | 23.34 | |
| 13 | 10 | 34.22 | |
| 14 | | | |

Here are the steps used to add a trendline to a chart:

1. Create a worksheet containing the data in Figure 5.45.
2. Create an XY Scatter chart from the data.
3. Add a chart layout with titles to the chart, as shown in Figure 5.46.

**Figure 5.46**
The algae growth data plotted on an XY Scatter chart.

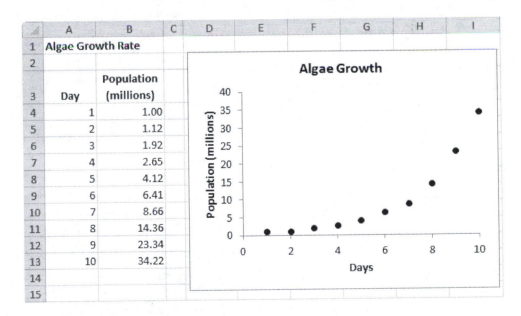

4. Right-click on the data series (right-click on any marker).
5. Select **Add Trendline . . .** from the pop-up menu. The Format Trendline dialog box (Figure 5.47) will open.

6. Select the following options (indicated in Figure 5.47):
   - Exponential Type
   - Display Equation on chart
   - Display *R*-squared value on chart
7. Click the **Close** button to close the Format Trendline dialog box.

The results are shown in Figure 5.48.

**Figure 5.47**
The Format Trendline
dialog box.

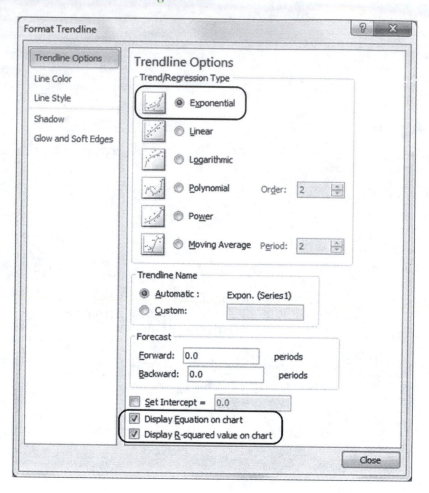

**Figure 5.48**
The algae growth data with
exponential trendline.

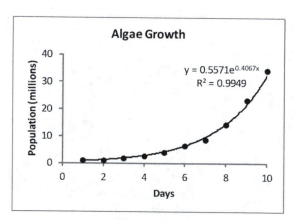

## PRACTICE!

Practice adding trendlines to charts and modifying trendline options.

First, create the worksheet of student score data shown in Figure 5.49. Then, create an XY Scatter chart of the data.

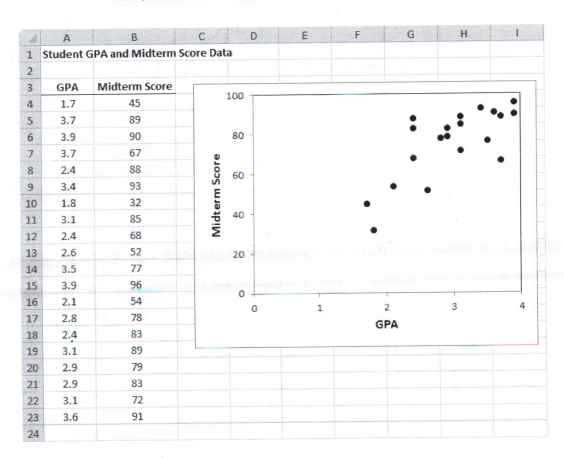

| | A | B |
|---|---|---|
| 1 | Student GPA and Midterm Score Data | |
| 2 | | |
| 3 | GPA | Midterm Score |
| 4 | 1.7 | 45 |
| 5 | 3.7 | 89 |
| 6 | 3.9 | 90 |
| 7 | 3.7 | 67 |
| 8 | 2.4 | 88 |
| 9 | 3.4 | 93 |
| 10 | 1.8 | 32 |
| 11 | 3.1 | 85 |
| 12 | 2.4 | 68 |
| 13 | 2.6 | 52 |
| 14 | 3.5 | 77 |
| 15 | 3.9 | 96 |
| 16 | 2.1 | 54 |
| 17 | 2.8 | 78 |
| 18 | 2.4 | 83 |
| 19 | 3.1 | 89 |
| 20 | 2.9 | 79 |
| 21 | 2.9 | 83 |
| 22 | 3.1 | 72 |
| 23 | 3.6 | 91 |
| 24 | | |

**Figure 5.49**
Student score data for trendline practice.

Add a linear trendline to the chart, as follows:

1. Right-click on the data series (right-click on any marker).
2. Select **Add Trendline . . .** from the pop-up menu. The Format Trendline dialog box will open.
3. Select the following options:
   o Linear Type
   o Set Intercept = 0
   o Display Equation on chart
   o Display R-squared value on chart
4. Forecast Backward 1.7 periods.
5. Click the **Close** button to close the Format Trendline dialog box.

Your graph should look something like Figure 5.50.

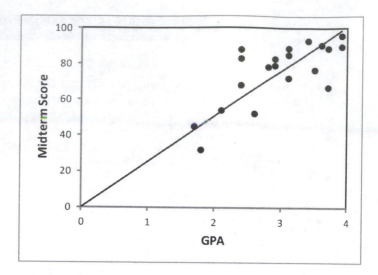

**Figure 5.50**
The student score data with linear trendline.

Next, try a polynomial trendline with a high order.

1. Right-click on the trendline.
2. Select **Format Trendline . . .** from the pop-up menu. The Format Trendline dialog box will open.
3. Select the following options:
   o Polynomial Type
   o Order set to 6
4. Click the **Close** button to close the Format Trendline dialog box.

What does the trendline look like when a high-order polynomial is used with noisy data?

Finally, use the moving average trendline.

1. Right-click on the trendline.
2. Select **Format Trendline . . .** from the pop-up menu. The Format Trendline dialog box will open.
3. Select the following options:
   o **Moving Average Type**
   o **Period set to 2**
4. Click the **Close** button to close the Format Trendline dialog box.

What does the moving average trendline look like when the data is not sorted before plotting?

## 5.8 USING THE GOAL SEEK TOOL

The next two sections introduce a technique for solving complex equations called *iterative solutions*. Iterative solutions use an initial guess for each variable in the equation and then solve both sides of the equation. Unless you are very lucky, the two sides of the equation evaluated by using the guess will not be equal. If they are not equal, another guess value is tried. This process continues until a value is found that satisfies the equation, or until the allowed number of guesses has been met.

The trick to getting solutions by using iterative methods is coming up with good guesses. The next two sections present different methods. The *Goal Seek* tool (this section) uses a simple method that is easy to use, but may fail to find a solution. The *Solver* (section 5.9) uses a more powerful technique to choose guesses and can usually find a solution, but it is a bit harder to learn to use.

The Goal Seek tool is used to find the input values of an equation when the results are known. It takes an initial guess for an input value and uses iterative refinement to attempt to locate the real input value.

As an example, we will use the Goal Seek tool to find the solution of a polynomial equation:

$$f(x) = 3x^3 + 2x^2 + 4 = 0.$$

The equation $f(x)$ has a real solution, which is approximately $x = -3$. We can use the initial guess of $x = 1$ to start the Goal Seek process, which will then attempt to converge on a more accurate value for $x$.

> **Caution!** The Goal Seek tool does not always converge. And, with some functions, the initial guess must closely approximate the real solution if Goal Seek is to converge.

To see how the Goal Seek works, use the following steps:

1. Create a worksheet that resembles Figure 5.51. Note the coding for $f(x)$ in the Formula box. The formula should be placed in cell B3.
2. Place 1, which is our initial guess for $x$, in cell B4. This results in $f(x)$ evaluating to a value of 9, which is displayed in cell B3.
3. Open the Goal Seek dialog box (shown in Figure 5.52) using these Ribbon options: **Data tab → Data Tools group → What-If Analysis drop-down menu → Goal Seek . . . button.** (Excel 2003: Tools → Goal Seek.)
4. Set the Goal Seek fields as follows:
   a. Set cell → B3 (this is the *target cell*, or the goal of the goal seek operation).
   b. To value → 0 (this is the result value that we want Goal Seek to find).
   c. By changing cell → B4 (this is the cell where the value is changed to try to meet the goal).
5. Click **OK** to start the Goal Seek operation.

The Goal Seek tool will try many values in cell B4, trying to meet the goal of a value of 0 in cell B3. The Goal Seek tool might find a pretty good answer and quit, or it could stop after a large number of attempts. In Excel 2007 the Goal Seek Status box (Figure 5.53) tells you whether or not a solution was found. In our example, a solution was found. The result is shown in Figure 5.54. Goal Seek found that a value of $x = -1.37347$ produces a solution, $f(x) = 0.0000375$ which is pretty close to zero.

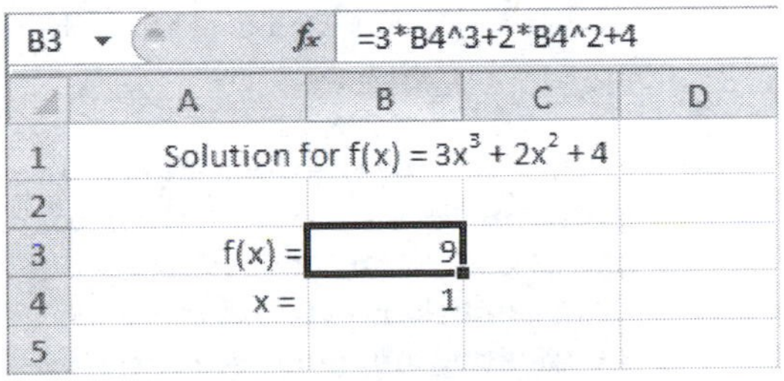

**Figure 5.51**

Solving a polynomial using the Goal Seek tool.

| B3 | ▼ | | $f_x$ | =3*B4^3+2*B4^2+4 | |
|---|---|---|---|---|---|
| | A | | B | C | D |
| 1 | Solution for f(x) = 3x³ + 2x² + 4 | | | | |
| 2 | | | | | |
| 3 | f(x) = | | 9 | | |
| 4 | x = | | 1 | | |
| 5 | | | | | |

**Figure 5.52**
The Goal Seek dialog box in use.

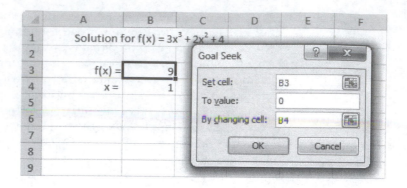

**Figure 5.53**
The Goal Seek Status box.

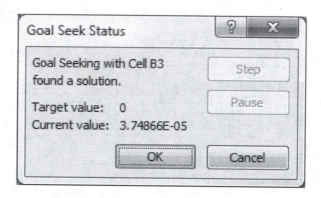

**Figure 5.54**
The solution after using the Goal Seek tool.

| | A | B | C | D |
|---|---|---|---|---|
| 1 | Solution for $f(x) = 3x^3 + 2x^2 + 4$ | | | |
| 2 | | | | |
| 3 | $f(x) =$ | 3.75E-05 | | |
| 4 | $x =$ | -1.37347 | | |
| 5 | | | | |

## 5.9 USING SOLVER FOR OPTIMIZATION PROBLEMS

Excel's *Solver* is designed to solve constrained non-linear optimization problems. Solving these complex problems is an important task for engineers in petroleum, defense, financial, agricultural, and process-control industries.

While the Solver is capable of handling complex optimization problems, you can also use it for more mundane problems. We will take a look at constrained optimization problems later in this section, but first let's introduce the Solver by using the same polynomial that was solved by using the Goal Seek tool in the last section:

$$f(x) = 3x^3 + 2x^2 + 4 = 0.$$

The Solver is an Excel add-in that must be installed and activated before it can be used. To see if Solver is active on your system, look on the Ribbon for a **Solver Button**: Data tab → Analysis group → Solver button. (Excel 2003: Tools → Solver.) If the button is present, the Solver is ready to use. If not, the Solver must either be activated (common) or installed (uncommon).

### 5.9.1 Activating the Excel Solver

If the **Solver** button appears on the Ribbon's Data tab, you can skip this step.

In Excel 2007 and 2010, add-ins like the Solver are managed using the **Add-Ins** panel on the Excel Options dialog box. To access the Excel Options dialog box, use the following options:

- Excel 2010: **File tab → Options button**
- Excel 2007: **Office button → Excel Options button**

The Excel Options dialog, shown in Figure 5.55, will open. Use the **Add-Ins** panel to activate the Solver.

**Figure 5.55**

The Add-Ins panel on the Excel Options dialog.

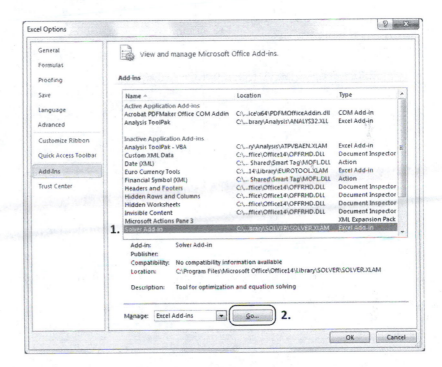

Look for "Solver Add-In" in the **Inactive Application Add-Ins** list (highlighted in Figure 5.55).

- If "Solver Add-In" is not listed in the **Inactive Application Add-Ins** list, it must be installed from the Excel program CDs (this is not common).
- If "Solver Add-In" is listed in the **Inactive Application Add-Ins** list, the Solver has been installed and you simply need to activate it.

The Solver can be activated with the following steps:

1. Select "Solver Add-In" in the **Inactive Application Add-Ins** list.
2. Click **Go . . .** to open the Add-Ins dialog box, shown in Figure 5.56.
3. Check the box labeled **Solver Add-In** (as illustrated in Figure 5.56).
4. Click **OK** to close the Add-Ins dialog box.
5. Click **OK** to close the Excel Options dialog box.

The Solver is now active on your system.

(In Excel 2003, open the Add-Ins dialog box using menu options: Tools → Add-Ins, then check the box labeled **Solver Add-In** (or, on older systems, just "Solver") as illustrated in Figure 5.56.)

**Figure 5.56**

The Add-Ins dialog box.

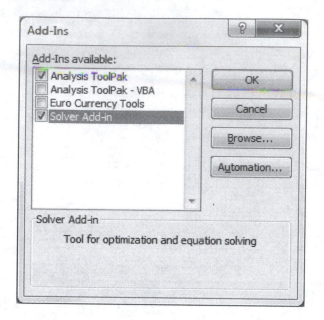

## 5.9.2 Using the Solver

To find the real solution to the example polynomial

$$f(x) = 3x^3 + 2x^2 + 4 = 0,$$

we first set up the same worksheet used in Section 5.8. This is shown in Figure 5.57.

**Figure 5.57**

Preparing to solve a polynomial using the Solver.

| B3 | ▼ | $f_x$ | =3*B4^3+2*B4^2+4 |
|---|---|---|---|

| | A | B | C | D |
|---|---|---|---|---|
| 1 | Solution for f(x) = $3x^3 + 2x^2 + 4$ | | | |
| 2 | | | | |
| 3 | f(x) = | 9 | | |
| 4 | x = | 1 | | |
| 5 | | | | |

Then, follow these steps to solve this problem using the Solver:

1. Open the Solver Parameters dialog box (shown in Figure 5.58) using these Ribbon options: **Data tab → Analysis group → Solver button**. (Excel 2003: Tools → Solver.)
2. Set the following Solver parameters:
   a. Set Target Cell → B3
   b. Equal to → Value of → 0
   c. By Changing Cells → B4
3. Click the **Solve** button to activate the Solver.

The Solver will try various values in cell B4, trying to find an $x$ value that makes $f(x) = 0$. It usually succeeds, and it tells you when it finds a solution by displaying the

Solver Results box shown in Figure 5.59. The result of using the Solver to solve the polynomial is shown in Figure 5.60. The solutions found with Goal Seek and the Solver are identical to at least 5 decimal places (shown in Figures 5.54 and 5.60).

**Figure 5.58**
The Solver Parameters dialog box.

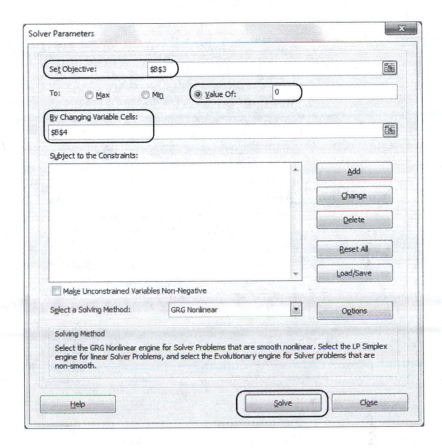

**Figure 5.59**
The Solver Results box.

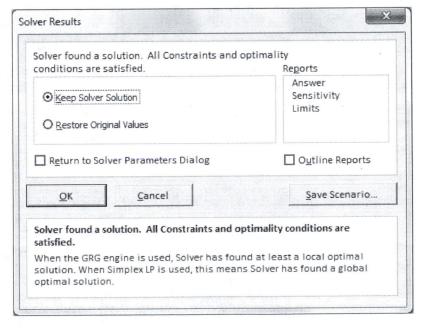

**Figure 5.60**
The solution found with the Solver.

| ⁄ | A | B | C | D |
|---|---|---|---|---|
| 1 | Solution for f(x) = 3x³ + 2x² + 4 | | | |
| 2 | | | | |
| 3 | f(x) = | 3.04E-07 | | |
| 4 | x = | -1.37347 | | |
| 5 | | | | |

### 5.9.3 Setting Up an Optimization Problem in Excel

What is an optimization problem? Many engineering problems have more than one solution, so engineers look for the best, or optimal, solution.

Engineers choose among a range of possible solutions by first specifying any limits to the input parameters of the problem (availability of raw materials or labor, for example). Given the limitations, the problem becomes one of finding the best solution, which in mathematical terms is a minimum or maximum solution. Engineers often seek solutions that maximize the strength of a load-bearing structure subject to a weight limit, or designs that minimize the weight of an aircraft while making sure it can handle the stresses of flight and landing.

The equation that is to be maximized (or minimized) is called the *objective function*. The limitations to the input parameters of the objective function are called *constraints*. Problems of this type—finding a minimum or maximum, given multiple constraints—are called *optimization problems.*

The most difficult part of solving an optimization problem is setting up the objective function and identifying the constraints. An objective function takes the form

$$y = f(x_1, x_2, \dots, x_n).$$

The independent variables ($x$) are limited by $m$ constraints, which take the form

$$c_i(x_1, x_2, \dots, x_n) = 0 \text{ for } i = 1, 2, \dots, m.$$

The constraints may also be expressed as inequalities. Excel Solver can handle both linear and non-linear constraints. However, non-linear constraints must be continuous functions. Although the constraints are expressed as functions, they are always evaluated within a range of precision called the *tolerance*. A constraint such as $x_1 < 0$, if the tolerance is large enough, may be evaluated as TRUE when $x_1 = 0.0000003$.

Non-linear optimization problems may have multiple minima or maxima. In a minimization problem, all solutions, except the absolute minimum, are called *local minima*. The solution that is chosen by Solver is dependent on the initial starting point for the solution. The initial guess should be as close to the real solution as possible. These are problems with optimization in general, not just with Excel Solver.

This has probably been rather confusing. A couple of examples will make things clearer. Let's proceed by setting up examples of both a linear and a non-linear optimization problem.

### 5.9.4 Linear Optimization Example

Assume that you wish to maximize the profit for producing widgets. The widgets come in two models: economy and deluxe. The economy model sells for $49.00, and the deluxe model sells for $79.00. The cost of production is determined primarily by labor costs, which are $14.00 per hour. The union limits the workers to a total of 2,000 hours per month. The economy widget can be built in 3 person-hours and the deluxe widget can be built in 4 person-hours. The management believes that it can sell up to 600 deluxe widgets per month and up to 1,200 economy widgets. Since you have a limited workforce, the main variable under your control is the ability to balance the number of economy units versus deluxe units that are built. Your job is to determine how many economy widgets and how many deluxe widgets should be built to maximize the company's profit.

The independent variables are as follows:

$$w_1 = \text{the number of economy widgets produced each month};$$

$$w_2 = \text{the number of deluxe widgets produced each month}.$$

The target that you wish to maximize is the profit, $p$. It is described mathematically by the following objective function:

$$p = (\$49 - (3 \text{ person-hours} * \$14/\text{hour}))w_1$$
$$+ (\$49 - (3 \text{ person-hours} * \$14/\text{hour}))w_2$$
$$= 7w_1 + 23w_2$$

The constraints can be expressed mathematically as limitations on $w_1$ and $w_2$. The maximum number of widgets to be produced is limited by both sales and labor availability. The sales limitations (imposed by management) can be expressed as

$$w_1 \leq 1{,}200 \text{ widgets}$$

and

$$w_2 \leq 600 \text{ widgets}.$$

The availability of labor is limited by the labor union and can be expressed as

$$3w_1 + 4w_2 \leq 2{,}000.$$

Finally, the general constraint of non-negativity is imposed on $w_1$ and $w_2$, since you cannot produce a negative number of widgets:

$$w_1 + w_2 \geq 0.$$

Figure 5.61 shows how to set up the widget problem in a worksheet. Cells C4 and C5 have been named *Weconomy* and *Wdeluxe*, respectively, to make the formulas more readable. The formulas in cells C8 and C11 are displayed in cells D8 and D11.

**Figure 5.61**
Worksheet for linear optimization example.

| | A | B | C | D | E |
|---|---|---|---|---|---|
| 1 | Widget Profit Optimization Worksheet | | | | |
| 2 | | | | | |
| 3 | Independent Variables | | | | |
| 4 | | Economy Widgets: | 0 | units per month | |
| 5 | | Deluxe Widgets: | 0 | units per month | |
| 6 | | | | | |
| 7 | Objective Function | | | | |
| 8 | | Profit, P: | 0 | =7*Weconomy+23*Wdeluxe | |
| 9 | | | | | |
| 10 | Constraints | | | | |
| 11 | | Labor Constraint: | 0 | =3*Weconomy+4*Wdeluxe | |
| 12 | | | | | |

You have already completed the hardest part of using the Solver, which is setting up the worksheet. To run the Solver, perform these steps:

1. Open the Solver Parameters dialog box (shown in Figure 5.62) using these Ribbon options: **Data tab → Analysis group → Solver button**. (Excel 2003: Tools → Solver.)

2. Set the following Solver parameters:
   ○ Set Target Cell → C8 (The cell holding the objective function, aka, the profit function.)
   ○ Equal to → Max (We want to maximize the profit.)
   ○ By Changing Cells → C4:C5 (The number of each type of widget.)
3. Add each of the following five constraints (shown in Figure 5.62):
   ○ C11 <= 2000—this is the labor constraint
   ○ Weconomy <= 1200
   ○ Weconomy >= 0
   ○ Wdeluxe <= 600
   ○ Wdeluxe >=0
4. Click the **Options** button to open the Solver Options dialog box, shown in Figure 5.63.
5. The Solver in Excel 2010 automatically selects an appropriate solving method based on the characteristics of the problem. As indicated in Figure 5.62, the Excel 2010 Solver selected a **Simplex LP** (linear programming) solution method for this linear problem.
   ○ For earlier versions of the Solver (2003, 2007), check the **Assume Linear Model** box for this linear optimization problem.
6. Click **OK** to close the Solver Options dialog box.
7. Click the **Solve** button to activate the Solver.

The Solver Results dialog box will appear, as depicted in Figure 5.64.

**Figure 5.62**
The Solver Parameters for the linear optimization problem.

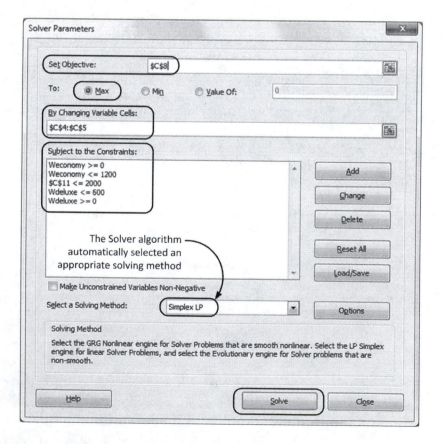

**Figure 5.63**
The Solver Options dialog box (set Assume Linear Model in pre-2010 Solvers).

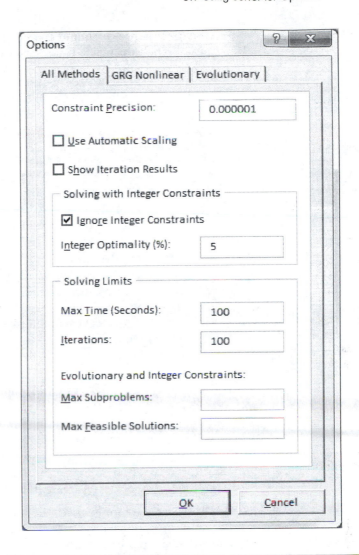

The Solver Results box should show that Solver has found a solution. There are three types of reports available from Solver: Answer, Sensitivity, and Limits.

1. Select all three types of reports.
2. Click **OK** to close the Solver Results box and see the solution.

**Figure 5.65**
The optimized solution.

| | A | B | C | D | E |
|---|---|---|---|---|---|
| 1 | Widget Profit Optimization Worksheet | | | | |
| 2 | | | | | |
| 3 | Independent Variables | | | | |
| 4 | | Economy Widgets: | 0 | units per month | |
| 5 | | Deluxe Widgets: | 500 | units per month | |
| 6 | | | | | |
| 7 | Objective Function | | | | |
| 8 | | Profit, P: | 11500 | =7*Weconomy+23*Wdeluxe | |
| 9 | | | | | |
| 10 | Constraints | | | | |
| 11 | | Labor Constraint: | 2000 | =3*Weconomy+4*Wdeluxe | |
| 12 | | | | | |

The worksheet will now contain modified values for the input parameters and objective function, as shown in Figure 5.65. The optimized worksheet shows that the maximum profit, $11,500 per month, is achieved by producing only deluxe widgets. Only 500 deluxe widgets can be produced per month, but the company can sell 600 per month; thus, the available labor pool is a limiting constraint.

In addition, the Solver will place three new worksheets in the workbook:

- The **Answer report** summarizes the initial and final values of the input parameters and the optimized variable.
- The **Sensitivity report** describes information about the marginal effects of making small changes in the constraints. Sometimes, a small constraint change can make a large difference in the output. For non-linear models, these are called *Lagrange multipliers*. For linear models, these are called either *dual values* or *shadow prices*.
- The **Limits report** shows the effect on the solution as each input parameter is set to its minimum or maximum limit.

As manager, you can easily modify the constraints and rerun Solver to see how the optimal solution might change. You can rapidly observe the effect on profit of:

- hiring more laborers
- modifying prices
- adjusting the widget mix

### 5.9.5 Non-Linear Optimization Example

As an example of non-linear optimization, we will use an optimization problem for which the solution is obvious. This will familiarize you with the process of setting up a non-linear optimization problem and convince you that the results are correct.

The objective function that we wish to minimize is

$$y = 100(x_2 - x_1^2)^2 + (1 - x_1)^2,$$

with the non-negativity constraints

$$x_1 \geq 0$$

and

$$x_2 \geq 0.$$

Since the terms $(x_2 - x_1^2)^2$ and $(1 - x_1)^2$ must be positive for real numbers $x_1$ and $x_2$ we know the answer. The minimum $y$ is zero when $x_1 = 1$ and $x_2 = 1$. The worksheet for this example is shown in Figure 5.66.

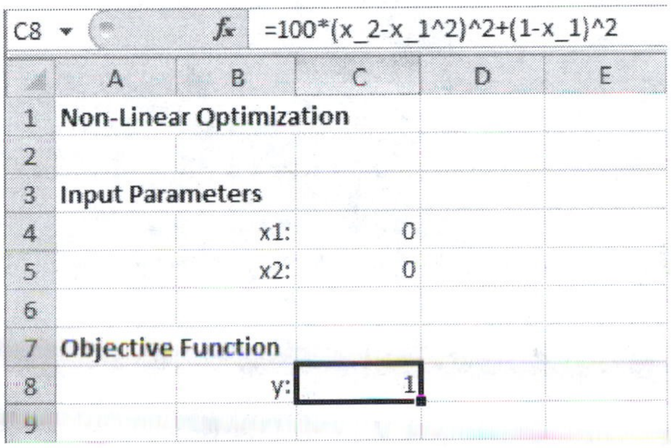

**Figure 5.66**

Worksheet for non-linear optimization example.

Complete the following steps to use the Solver to perform non-linear optimization:

1. Open the Solver Parameters dialog box (shown in Figure 5.67) using these Ribbon options: **Data tab → Analysis group → Solver button**. (Excel 2003: Tools → Solver.)
2. Set the following Solver parameters:
   **a.** Set Target Cell → C8 (The cell holding the objective function.)
   **b.** Equal to → Min (We want to minimize the function.)
   **c.** By Changing Cells → C4:C5 ($x_1$ and $x_2$)
3. Add each of the following non-negativity constraints (shown in Figure 5.67):
   **a.** x_1 >= 0
   **b.** x_2 >=0
4. Click the **Options** button to open the Solver Options dialog box.
5. Make sure the **Assume Linear Model** box (Excel 2003 and 2007) is not checked.
6. Click **OK** to close the Solver Options dialog box.
7. Click the **Solve** button to activate the Solver.

The result is shown in Figure 5.68; the results produced by Solver are good approximations of the true minimum.

There are other local minima for this objective function. Try setting the initial parameters to $x_1 = 5$ and $x_2 = 3$ and rerunning the Solver. The results produced by Solver (Figure 5.69) show that the algorithm is stuck in a local minimum.

**Figure 5.67**
The Solver Parameters dialog box for the non-linear optimization example.

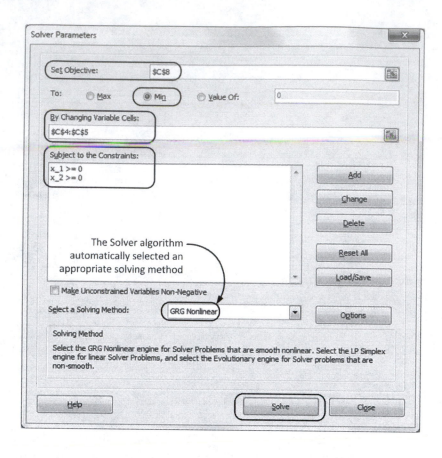

**Figure 5.68**
The optimized solution.

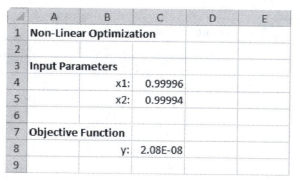

|  | A | B | C | D | E |
|---|---|---|---|---|---|
| 1 | **Non-Linear Optimization** |  |  |  |  |
| 2 |  |  |  |  |  |
| 3 | **Input Parameters** |  |  |  |  |
| 4 |  | x1: | 0.99996 |  |  |
| 5 |  | x2: | 0.99994 |  |  |
| 6 |  |  |  |  |  |
| 7 | **Objective Function** |  |  |  |  |
| 8 |  | y: | 2.08E-08 |  |  |
| 9 |  |  |  |  |  |

**Figure 5.69**
Solver results when started with $x_1 = 5$ and $x_2 = 3$.

|  | A | B | C | D | E |
|---|---|---|---|---|---|
| 1 | **Non-Linear Optimization** |  |  |  |  |
| 2 |  |  |  |  |  |
| 3 | **Input Parameters** |  |  |  |  |
| 4 |  | x1: | 0.630629 |  |  |
| 5 |  | x2: | 0.399948 |  |  |
| 6 |  |  |  |  |  |
| 7 | **Objective Function** |  |  |  |  |
| 8 |  | y: | 0.136943 |  |  |
| 9 |  |  |  |  |  |

APPLICATION

## YIELD STRENGTH OF MATERIALS

Materials science is an important field of study for engineers that covers the electronic, optical, mechanical, chemical, and magnetic properties of metals, polymers, composite materials, and ceramics.

Crystalline materials are subject to *slip deformation* when a shear stress is applied to the material. The deformation occurs when atomic planes slide along the directions of densest atomic packing, as shown in Figure 5.70.

**Figure 5.70**

Slip deformation.

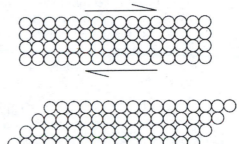

When crystalline materials, such as metals and alloys, are formed by cooling molten metal, separate crystals form in the melt and grow together. The boundaries between the growing crystals form barriers to slip deformation, increasing the observed *yield strength* of the metal. The relationship between grain size (i.e., the size of the individual crystals in the metal) and observed yield strength, $\sigma_y$, is described by the Hall–Petch equation,

$$\sigma_y = \sigma_o + k_y \frac{1}{\sqrt{d}},$$

where $\sigma_o$ is the yield strength of the pure metal (i.e., single crystal, $d = \infty$). The value of the proportionality factor, $k_y$, depends on the material and can be obtained by regression analysis from experimental data (see Figure 5.71).

For the carbon-steel data shown here, the coefficients for the Hall–Petch equation can be found by using Excel:

$$\sigma_y = 60.466 \, \frac{MN}{m^2} + 689.49 \, \frac{MN \, \mu m^{0.5}}{m^2} \frac{1}{\sqrt{d}}.$$

The value would typically be reported as

$$k_y = 0.689 \, \frac{MN}{m^{3/2}}.$$

**How Did They Do That?**

The Hall–Petch equation is an example of an equation that can be written in linear form

$$y = ax + b,$$

where $a$ is the slope of the line through the data values and $b$ is the $y$-intercept. Comparing terms with the Hall–Petch equation, you see that

$$\sigma_y = y;$$

$$\sigma_o = b;$$

$$\frac{1}{\sqrt{d}} = x.$$

A plot of $\frac{1}{\sqrt{d}}$ on the *x*-axis against $\sigma_y$ on the *y*-axis shows the desired linear relationship. We can use an Excel trendline determine the equation of the line through the data, using the following steps:

1. Create a worksheet by using the carbon-steel data shown in Figure 5.71.
2. Add a third column to your worksheet that contains 1/SQRT(d).
3. Create the XY Scatter chart like the one shown in Figure 5.71.
4. Right-click on one of the data points and select **Add Trendline . . .** from the pop-up menu. The Format Trendline dialog box will open.
5. Choose a **Linear** trendline and check the box labeled **Display equation on chart**.
6. Click the **Close** button to close the Format Trendline dialog box.

Excel will determine the regression fit through the data points and display the equation of the regression line on the graph, as shown in Figure 5.71.

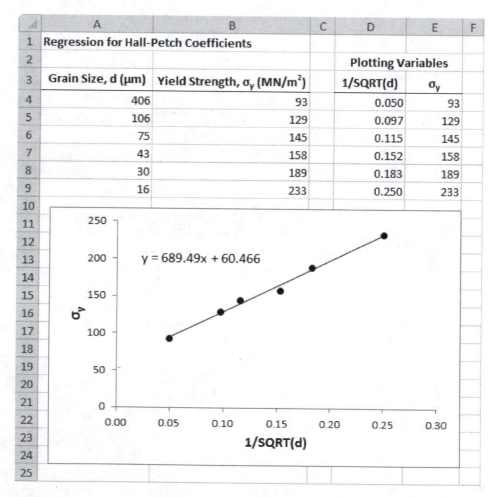

**Figure 5.71**
Table of carbon-steel data and results of regression analysis.

## KEY TERMS

| | | |
|---|---|---|
| Analysis ToolPak | *GROWTH* function | objective function |
| bin | growth trend | optimization |
| constraint | histogram | regression |
| correlation | iterative methods | Solver |
| correlation coefficient, *r* | linear regression | standard deviation |
| data analysis | linear trend | statistics |
| descriptive statistics | mean | trend analysis |
| exponential growth | median | *TREND* function |
| frequency | mode | trendline |
| Goal Seek | multiple regression | |

# SUMMARY

### Creating a Histogram

- Open the Data Analysis dialog box: **Data tab** ➔ **Analysis group** ➔ **Data Analysis button**.
- Select **Histogram** from the list of data analysis tools.
- Click **OK** to close the Data Analysis dialog box and open the Histogram dialog box.
- Set the **Input Range**.
- Set the **Bin Range**. (Excel will assign default bin values if you skip this step.)
- Select the output location.
- Request the histogram chart by checking the **Chart Output** box.
- Click OK to close the Histogram dialog box and create the histogram chart.

### Calculating Descriptive Statistics

1. Open the Data Analysis dialog box: **Data tab** ➔ **Analysis group** ➔ **Data Analysis button**.
2. Select **Descriptive Statistics** from the list of data analysis tools.
3. Click **OK** to close the Data Analysis dialog box and open the Descriptive Statistics dialog box.
4. Select the data values as the **Input Range** (shown in Figure 5.18).
5. Check the **Labels in First Row** box if you data includes a column heading.
6. Choose a location for the output table.
7. Check the **Summary statistics** box.
8. Click **OK** to close the Descriptive Statistics dialog box and calculate the statistic values.

| Statistic | Description |
|---|---|
| Mean | The data set shows scatter around a central point, which is called the *mean*. |
| Standard Error | The *standard error* is used to estimate how accurately the *sample mean* represents the *population mean*. |
| Median | The *median* is the middle value of the sorted data set. |
| Mode | The *mode* is the most commonly occurring value in the data set. |

| Statistic | Description |
|---|---|
| Standard Deviation | The *standard deviation* is a measure of the data's variability. |
| Sample Variance | The *sample variance* is the square of the *sample standard deviation*. |
| Kurtosis | The *kurtosis* provides information about the "peakedness" of the data distribution compared with a normal distribution. |
| Skewness | The *skewness* provides information about the extent of asymmetry of the distribution about its *mean*. |
| Range | The *range* is the difference between the maximum and minimum values in the data set. |
| Minimum | *Minimum* indicates the smallest value in the data set. |
| Maximum | *Maximum* indicates the largest value in the data set. |
| Sum | The *sum* is the result of adding all values in the data set. |
| Count | The *count* is the number of values in the data set. |

### Computing a Correlation

1. Open the Data Analysis dialog box: **Data tab → Analysis group → Data Analysis button**.
2. Select **Correlation** from the list of data analysis tools.
3. Click **OK** to close the Data Analysis dialog box and open the Correlation dialog box.
4. Set the **Input Range** for the calculation.
5. Check the **Labels in First Row** box if your data set included a column heading.
6. Choose a location for the output table.
7. Click **OK** to close the Correlation dialog box and calculate the *correlation coefficient, r.*

### Performing a Linear Regression

Two ways to perform a linear regression in Excel include:

1. Adding a trendline (except for the moving average type) to a chart.
2. Using the Regression tool in the Analysis ToolPak.

#### Adding a Trendline to a Chart

1. Create an XY Scatter chart from your data.
2. Right-click on the data series.
3. Select **Add Trendline . . .** from the pop-up menu. The Format Trendline dialog box will open.
4. Select the type of trendline desired.
5. Set desired options:
   o Set intercept value
   o Display Equation on chart
   o Display R-squared value on chart
6. Click the Close button to close the Format Trendline dialog box and create the trendline.

*Types of Trendlines*

| Trendline Type | Formula |
| --- | --- |
| Linear | Calculates the least squares fit by using $y = mx + b$. ($m$ is the slope and $b$ is a constant.) |
| Logarithmic | Calculates least squares by using $y = c \cdot \ln(x) + b$. ($c$ and $b$ are constants.) |
| Polynomial | Calculates least squares for a line by using $y = b + c_1x + c_2x^2 + \cdots + c_nx^n$. |
| | ($b, c_1, c_2, \ldots, c_n$ are constants, the order can be set in the Add Trendline dialog box, and the maximum order is 6.) |
| Exponential | Calculates least squares by using $y = ce^{bx}$. ($c$ and $b$ are constants.) |
| Power | Calculates least squares by using $y = cb^x$. ($c$ and $b$ are constants.) |
| Moving Average | Not a regression trendline. Calculates the series of moving averages by using $$F_{(t+1)} = \frac{1}{N}\sum_1^N A_{t-j+1}.$$ $N$ is the number of prior periods to average, $A_t$ is the value at time $t$, $F_t$ is the forecasted value at time $t$. |

### Using the Regression Tool in the Analysis ToolPak

1. Open the Data Analysis dialog box: **Data tab → Analysis group → Data Analysis button**.
2. Select **Regression** from the list of data analysis tools.
3. Click **OK** to close the Data Analysis dialog box and open the Regression dialog box.
4. Set the **Input Y Range** for the regression analysis.
5. Set the **Input X Range** for the regression analysis.
6. Check the **Labels** box if you included column (or row) headings in the input ranges.
7. Choose a location for the output table.
8. Click **OK** to close the Regression dialog box and perform the regression analysis.

### Trend Analysis with the Series Dialog Box

*Linear Extension (preserves known values)*

1. Select known data points and the unknown (to be calculated) points.
2. Open the Series dialog box: **Home tab → Editing group → Fill drop-down menu → Series . . . button**.
3. Select **Rows** or **Columns** as appropriate for your data.
4. Select the **AutoFill** type.
5. Clear the **Trend** box.
6. Click **OK** to calculate the series values.

*Linear Replacement (replaces known values with calculated trend values)*

1. Select known data points and the unknown (to be calculated) points.
2. Open the Series dialog box: **Home tab → Editing group → Fill drop-down menu → Series . . . button**.
3. Select **Rows** or **Columns** as appropriate for your data.

4. Select the **Linear** type.

5. Check the **Trend** box.

6. Click **OK** to calculate the series values.

*Exponential Growth Replacement (replaces known values with calculated trend values)*

1. Select known data points and the unknown (to be calculated) points.

2. Open the Series dialog box: **Home tab → Editing group → Fill drop-down menu → Series . . . button**.

3. Select **Rows** or **Columns** as appropriate for your data.

4. Select the **Growth** type.

5. Check the **Trend** box.

6. Click **OK** to calculate the series values.

*Using the TREND Function (preserves known values)*

1. Select the unknown (to be calculated) points only.

2. Choose the **Insert Function** button on the Formula bar and select the Statistical category.

3. Choose the *TREND* function from the function list. The Function Arguments dialog box will appear.

4. Enter the **Known_y's** cell range.

5. Enter the **Known_x's** cell range.

6. Enter the **New_x's** cell range.

7. Press **Ctrl + Shift + Enter** to enter the formula in multiple cells.

## Iterative Solutions

Excel provides two iterative solvers:

- Goal Seek—simple to use, can fail to find roots.
- Solver—a little more complicated, can handle constraints and optimization, finds most roots.

*Using Goal Seek*

1. Create a worksheet that provides a cell for an initial value and a cell for the calculated result (aka target cell). (There can be many additional calculations between the initial value and the target cell, but Goal Seek uses two cells.)

2. Enter an initial guess.

3. Open the Goal Seek dialog box: **Data tab → Data Tools group → What-If Analysis drop-down menu → Goal Seek . . . button**.

4. Set the Goal Seek fields as follows:
   a. Set cell → (This is the *target cell*, or the goal of the goal seek operation.)
   b. To value → (This is the result value that we want Goal Seek to find.)
   c. By changing cell → (This is the cell where the value is changed to try to meet the goal.)

5. Click **OK** to start the Goal Seek operation.

*Using the Solver*

1. Create a worksheet that provides one or more cells for initial values, and a cell for the calculated result (aka target cell). (There can be many additional

calculations between the initial value(s) and the target cell, but the Solver requires at least two cells.)

2. Open the Solver Parameters dialog box: **Data tab → Analysis group → Solver button**.

3. Set the following Solver parameters:
   a. Set Target Cell
   b. Equal to → Options:
      i   Value of → <value>
      ii  Max
      iii Min
   c. By Changing Cells (The Solver will vary the values in multiple cells, if needed.)

4. Click the **Solve** button to activate the Solver.

## PROBLEMS

1.  Use the Analysis ToolPak Histogram feature and the data in Figure 5.6 to create a cumulative percentage histogram of the students' midterm exam grades. What percent of the students earned a score of 80 or lower on the midterm exam?

2.  Use the Descriptive Statistics selection from the Analysis ToolPak to find the mean and standard deviation for the student GPAs in Figure 5.6. Compute how many GPAs are within one standard deviation of the mean. Compute this by adding the standard deviation to the mean and then sub-tracting the standard deviation from the mean. How many GPAs lie within the range of the two numbers? What percentage of the GPAs lie within one standard deviation of the mean?

3.  Look at the traffic study data in Table 5.4. Perform a regression analysis for intersection 3. What is the linear equation that best fits the data? Use your linear equation to predict the traffic flow at this intersection in 2003.

4.  Generate a Scatter plot for the 11% column in Figure 5.72. Add an expo-nential growth trendline to the chart, then using the Format Trendline

**Table 5.4 Average Daily Traffic Flow at Four Downtown Intersections**

| | Average Daily Traffic Flow ($\times 1{,}000$) | | | |
| | Intersection # | | | |
| Year | 1 | 2 | 3 | 4 |
|---|---|---|---|---|
| 1996 | 25.3 | 12.2 | 34.8 | 45.3 |
| 1997 | 26.3 | 14.5 | 36.9 | 48.7 |
| 1998 | 28.6 | 14.9 | 42.6 | 43.2 |
| 1999 | 29.0 | 16.8 | 50.6 | 46.9 |
| 2000 | 32.4 | 17.6 | 70.8 | 54.9 |
| 2001 | 34.8 | 17.9 | 82.3 | 60.9 |

dialog box, forecast forward 20 periods (20 more years). By looking at the projected trendline, determine John's approximate nest egg at age 68 if he invests $10,000 at age 18 (assuming he hasn't started spending it).

**Figure 5.72**

Single-payment compound amount factors.

| | A | B | C | D | E | F | G | H | I |
|---|---|---|---|---|---|---|---|---|---|
| 1 | Compound Amount Factors | | | | | | | | |
| 2 | | | | | | | | | |
| 3 | Interest Rate: | | 6% | 7% | 8% | 9% | 10% | 11% | |
| 4 | | | | | | | | | |
| 5 | | Year | | | | | | | |
| 6 | | 0 | 1.0000 | 1.0000 | 1.0000 | 1.0000 | 1.0000 | 1.0000 | |
| 7 | | 10 | 1.7908 | 1.9672 | 2.1589 | 2.3674 | 2.5937 | 2.8394 | |
| 8 | | 20 | 3.2071 | 3.8697 | 4.6610 | 5.6044 | 6.7275 | 8.0623 | |
| 9 | | 30 | 5.7435 | 7.6123 | 10.0627 | 13.2677 | 17.4494 | 22.8923 | |
| 10 | | | | | | | | | |

5. Using an XY Scatter chart, graph the function

$$f(x) = x^3 + \sin(x/2) + 2x - 4,$$

for $x = [-2, 2]$ in increments of 0.1. Add a trendline to the chart. What type of trendline best fits the data? Display the equation and $R^2$ value on the chart.

6. Evaluate the function

$$y = Ln(2x) + \sin(x),$$

for $x = [1, 10]$ in increments of 0.1. Chart the results using an XY Scatter chart. From looking at the plot, what do you think is the minimum of the function in this range?

7. Compute values of

$$f(x) = 2x^3 - 13x - 9,$$

for $x = [-3, 3]$ in increments of 0.1. Plot the function by using an XY Scatter plot. You can see from the plot that $f(x) = 0$ near $x = -2.1, -0.8$, and 2.8. Use these three guesses and the Goal Seek tool to find more accurate solutions for $f(x) = 0$.

8. Use the Solver to minimize the objective function

$$f(x) = (x_1 + 2x_2 - 7)^2 + (2x_1 + x_2 - 5)^2,$$

with the constraints $-10 \le x_1, x_2 \le 10$. What are the values of $x_1$ and $x_2$ when $f(x)$ is at its minimum?

9. A cylindrical chemical petroleum tank is to be built to hold 6.8 m$^3$ of hazardous waste. Your task is to design the tank in a cost-effective manner by minimizing its surface area. Ignore the thickness of the walls in your design. Recall that the surface area $S$ of a right-angled cylinder is

$$S = 2\pi r^2 + 2\pi rh$$

and the volume $V$ of a cylinder is

$$v = \pi r^2 h$$

Use the Solver to minimize $r$ and $h$.

10. Enthalpy and entropy data for saturated steam between 100 and 300°C are shown in Table 5.5. Prepare a chart of the enthalpy data as a function of temperature, and add a trendline to the chart that fits the data well. Report the equation of the trendline and the $R^2$ value.

**Table 5.5 Properties of Saturated Steam**

| Temp. (°C) | Enthalpy (kJ/kg) | Entropy (kJ/kg K) |
|---|---|---|
| 100 | 2680 | 7.4 |
| 150 | 2750 | 6.8 |
| 200 | 2790 | 6.4 |
| 250 | 2800 | 6.1 |
| 300 | 2750 | 5.7 |

 11. The viscosity of a liquid varies greatly as the temperature of the liquid changes. An engineer simulating a hydroelectric power plant must account for the effect of temperature on the viscosity of water in his simulation. Create a chart of the viscosity and temperature data listed in Table 5.6, and find a trendline that fits the data. Report the equation of the line and the $R^2$ value.

**Table 5.6 Viscosity of Water**

| Temp. (°C) | $\mu$ (cP) |
|---|---|
| 0 | 1.8 |
| 5 | 1.5 |
| 10 | 1.3 |
| 15 | 1.1 |
| 20 | 1.0 |
| 25 | 0.9 |
| 30 | 0.8 |
| 35 | 0.7 |
| 40 | 0.6 |

12. The values charted in Figures 5.21 and 5.22 were computed by using the following equations:

Figure 5.24, left panel

$$y_1 = x + \sin(x)$$

$$y_2 = 2 + 0.9x + \sin(x)$$

Figure 5.24, right panel

$$y_1 = x + \sin(x)$$

$$y_3 = 4 - 0.9x - \sin(x)$$

Figure 5.25

$$y_1 = x + \sin(x)$$

$$y_4 = 4 - 0.2x - 3\cos(x/3)$$

Use these equations to calculate $y$ values (i.e., $y_1$, $y_2$, $y_3$, $y_4$ values) for $x$ values between 0 and 5, with an interval of 0.2. Use these values with Excel's Correlation tool to determine the correlation coefficients for the following cases:

a. $y_1$, $y_2$—Figure 5.24, left panel
b. $y_1$, $y_3$—Figure 5.24, right panel
c. $y_1$, $y_4$—Figure 5.25

# 6

# Database Management Within Excel

## Sections

## Objectives

*After reading this chapter, you should be able to perform the following tasks:*

- Create a database within Excel.
- Enter data into an Excel database.
- Sort a database on one or more keys.
- Use filters to query databases.

## 6.1 INTRODUCTION

Microsoft Excel implements a rudimentary *database management system*[1] (DBMS) by treating lists in a worksheet as database records. This is helpful for organizing, *sorting*, and searching through worksheets that contain many related items. You can import

---

[1] A collection of programs that enables you to store, extract, and modify information from a database.

complete databases from external DBMS such as Microsoft Access, Oracle, dBase, and text files. You can create structured queries by using Microsoft Query that will retrieve selected information from external sources.

If you require a relational DBMS, then you are encouraged to use another, more complete software application such as Microsoft Access. However, the database functions within Excel are adequate for many problems. An example of one way that an engineer might use this functionality is to import experimental data that has been stored in a relational DBMS in order to perform analysis on the data, using Excel's built-in functions.

## 6.1.1 Database Terminology

A database within Excel is sometimes called a *list*. A **database** can be thought of as an electronic file cabinet that contains a number of folders. Each folder contains similar information for different objects. For example, each folder might contain the information about a student at a college of engineering. The database is the collection of all student folders.

The data in each folder are organized in a similar fashion. For example, each folder might include a student's:

- last name
- first name
- college identification number
- address
- department
- class
- and other pertinent information.

Using database terminology, each folder is called a **record**. Each data item is stored within a **field**. The title for each data item is called a **field name**.

Any region in an Excel worksheet can be defined as a database. Excel represents each record as a separate row. Each cell within the row is a field. The heading for each column is the field name.

Figure 6.1 depicts a small student database. Rows 3 through 12 each represent a student record. Each record has five fields. The field names are the column headings (e.g., **Last Name**).

**Figure 6.1**

Example of a student database.

| | A | B | C | D | E | F | G |
|---|---|---|---|---|---|---|---|
| 1 | | | | | | | |
| 2 | | **Last Name** | **First Name** | **GPA** | | **Department** | **Class** |
| 3 | | Phillips | Casey | 3.51 | | Electrical | Junior |
| 4 | | Noland | Rudy | 2.98 | | Chemical | Senior |
| 5 | | Murray | Susie | 3.92 | | Electrical | Senior |
| 6 | | Carter | Christine | 2.78 | | Civil | Junior |
| 7 | | Madison | Richard | 3.41 | | Mechanical | Junior |
| 8 | | Carter | Frank | 3.12 | | Chemical | Senior |
| 9 | | Kirk | William | 3.35 | | Civil | Sophomore |
| 10 | | Hahn | Eric | 3.87 | | Mechanical | Junior |
| 11 | | Garcia | Rob | 3.21 | | Electrical | Sophomore |
| 12 | | Ault | Linda | 3.78 | | Chemical | Senior |
| 13 | | | | | | | |

## 6.2 CREATING DATABASES

Most DBMSs store records in one or more separate files. The file delimits the boundaries of the database. Excel, however, stores a database as a region in a worksheet.

Excel must have some way of knowing where the database begins and ends in the worksheet. There are two methods for associating a region with a database.

- Leave a perimeter of blank cells around the database region.
- Explicitly name the region.

Because of the unique way that Excel delimits a database, the following tips are recommended:

- Maintain only one database per worksheet. This will speed up access to the sorting and filtering functions, and you will not need to name the database regions.
- Make sure that each column heading in the database is unique. If there were two headings for **Last Name**, for example, a logical query such as

  *Find all records with* **Last Name** *equal to* **Smith**

  would not make sense.

- Leave an empty column to the right of the database and an empty row at the bottom of the database (and to the left and top if the database does not include cell A1). Excel uses the empty row and column to mark the edge of the database. An alternative method is to assign a name to the region of the database. One disadvantage of assigning a name is that the allocated region may have to be redefined when records are added or deleted.
- Do not use cells to the right of the database for other purposes. Filtered rows may inadvertently hide these cells.

## 6.3 ENTERING DATA

Once the field names for the database have been created in the column headings, data may be entered by using several methods. One method of data entry is to type data directly into a cell. A database field may contain any legitimate Excel value, including numerical, date, text, or formula. For example, you might add a column to the database in Figure 6.1 that is titled **Full Name**. Instead of copying or retyping the first and last names of each student, the new field could concatenate the **First Name** and **Last Name** entries by using the following formula:

$$= \text{CONCATENATE}(B2, ", ", A2)$$

The result is shown in Figure 6.2.

A second method for entering data is to use a form. The ***data entry form*** is readily available from the menus in Excel 2003 (Data ➔ Form), but it is not part of the Ribbon. You can still use the form, but with Excel 2007 and 2010 you must first add the **Form . . .** button to the Quick Access Toolbar using the following steps:

1. Click the small down arrow next to the Quick Access Toolbar (illustrated in Figure 6.3).
2. Select **More Commands** from the Customize Quick Access Toolbar menu (see Figure 6.3). This opens the Excel Options dialog to the Customize panel as shown in Figure 6.4.
3. Select **Commands Not in the Ribbon** in the **Choose commands from** drop-down list.

**Figure 6.2**
Inserting a Full Name field.

| | A | B | C | D | E | F | G | H |
|---|---|---|---|---|---|---|---|---|
| | | | | fx | =CONCATENATE(B3,", ",C3) | | | |

| | A | B | C | D | E | F | G | H |
|---|---|---|---|---|---|---|---|---|
| 1 | | | | | | | | |
| 2 | | Last Name | First Name | Full Name | GPA | Department | Class | |
| 3 | | Phillips | Casey | Phillips, Casey | 3.51 | Electrical | Junior | |
| 4 | | Noland | Rudy | Noland, Rudy | 2.98 | Chemical | Senior | |
| 5 | | Murray | Susie | Murray, Susie | 3.92 | Electrical | Senior | |
| 6 | | Carter | Christine | Carter, Christine | 2.78 | Civil | Junior | |
| 7 | | Madison | Richard | Madison, Richard | 3.41 | Mechanical | Junior | |
| 8 | | Carter | Frank | Carter, Frank | 3.12 | Chemical | Senior | |
| 9 | | Kirk | William | Kirk, William | 3.35 | Civil | Sophomore | |
| 10 | | Hahn | Eric | Hahn, Eric | 3.87 | Mechanical | Junior | |
| 11 | | Garcia | Rob | Garcia, Rob | 3.21 | Electrical | Sophomore | |
| 12 | | Ault | Linda | Ault, Linda | 3.78 | Chemical | Senior | |
| 13 | | | | | | | | |

**Figure 6.3**
Adding a button to the
Quick Access Toolbar.

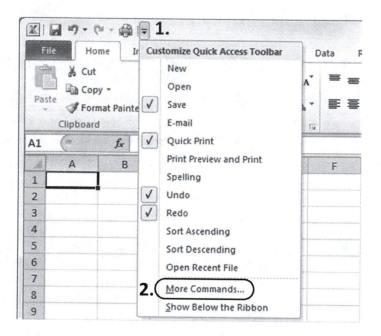

**Figure 6.4**
Adding the Form . . .
button to the Quick Access
Toolbar.

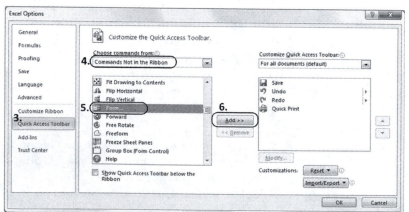

**Figure 6.5**
The **Form** button has been added to the Quick Access Toolbar.

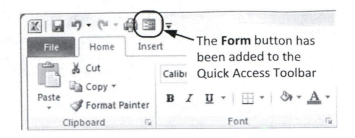

The **Form** button has been added to the Quick Access Toolbar

4. Select **Form . . .** from the list of available commands (left panel).
5. Click **Add** >>> to add the **Form . . .** button to the Quick Access Toolbar (shown in Figure 6.5).
6. Click **OK** to close the Excel Options dialog box.

Once the **Form . . .** button is available, you can use it for data entry with these steps:

1. Select any data cell in the database.
2. Click the **Form . . .** button.

The Data Entry form will appear, as depicted in Figure 6.6. The title of the Data Entry form will be the same as the name of the current worksheet, *Student DB* in this example.

**Figure 6.6**
The Form dialog. (Student Database is the name of the worksheet containing the database.)

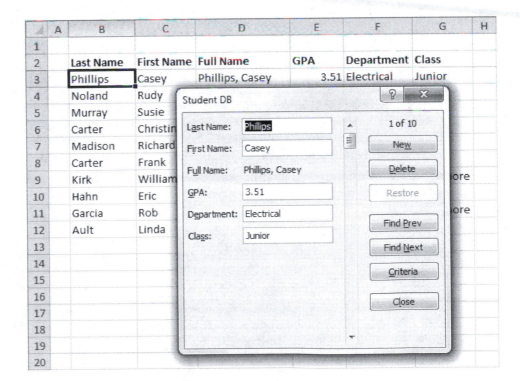

From the Data Entry form you can:

- Create a new record by clicking the **New** button.
- Scroll through the database with the slider control.
- Select adjacent records with the **Find Prev** and **Find Next** buttons.
- **Delete** records.
- Modify existing records.

## PRACTICE!

Before proceeding, it will be helpful for you to create the database depicted in Figure 6.1. This database will be used for the examples in the rest of the chapter. Practice entering some of the data by using the Data Entry form. Enter some of the data by typing directly into the worksheet. Which method is more resistant to typing errors?

## 6.4 SORTING A DATABASE

The power of a DBMS lies in its ability to search for information, rearrange data, and filter information.

To sort a database, follow this procedure:

1. Select any data cell in the database.
2. Open the Sort dialog box using any of these methods:
   a. From the Ribbon's Home tab: **Home tab → Editing group → Sort & Filter drop-down menu → Custom Sort . . . button**.
   b. From the Ribbon's Data tab: **Data tab → Sort & Filter group → Sort button**.
   c. Excel 2003: Data → Sort.

The Sort dialog box will appear, as depicted in Figure 6.7. The field on which the sort is made is called the *sort key*. Excel allows you to sort on multiple keys. In this example we will sort on **Last Name** and **First Name**.

3. Choose **Last Name** in A to Z (ascending) order as the first key.
4. Click **Add Level**, then choose **First Name** in A to Z order as the second key.
5. Make sure that the **My data has headers** box is checked.
6. Click **OK** to close the Sort dialog box and perform the sort.

The result is an alphabetical listing, by last name, of the student database as shown in Figure 6.8.

**CAUTION:** If you select a portion of the database before asking Excel to perform a sort, Excel will show a warning box that there are non-empty adjacent cells. Sorting a portion of the database will misalign the data in the various fields, and Excel tries to warn you not to do it. If you accidentally scramble the database, immediately click the **Undo** button on the Quick Access toolbar.

If you follow the procedure listed above and click in a single cell before performing any database operations, Excel will automatically select the entire database before sorting or filtering.

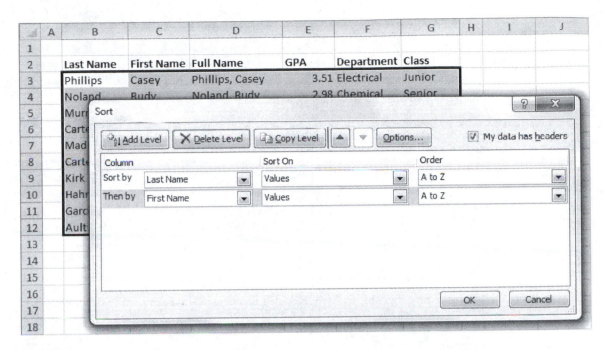

**Figure 6.7**
The Sort dialog box.

**Figure 6.8**
The sorted database.

| | Last Name | First Name | Full Name | GPA | Department | Class |
|---|---|---|---|---|---|---|
| 2 | Last Name | First Name | Full Name | GPA | Department | Class |
| 3 | Ault | Linda | Ault, Linda | 3.78 | Chemical | Senior |
| 4 | Carter | Christine | Carter, Christine | 2.78 | Civil | Junior |
| 5 | Carter | Frank | Carter, Frank | 3.12 | Chemical | Senior |
| 6 | Garcia | Rob | Garcia, Rob | 3.21 | Electrical | Sophomore |
| 7 | Hahn | Eric | Hahn, Eric | 3.87 | Mechanical | Junior |
| 8 | Kirk | William | Kirk, William | 3.35 | Civil | Sophomore |
| 9 | Madison | Richard | Madison, Richard | 3.41 | Mechanical | Junior |
| 10 | Murray | Susie | Murray, Susie | 3.92 | Electrical | Senior |
| 11 | Noland | Rudy | Noland, Rudy | 2.98 | Chemical | Senior |
| 12 | Phillips | Casey | Phillips, Casey | 3.51 | Electrical | Junior |

## PRACTICE!

Sort the student database in ascending order by Department and descending order by GPA within each department. Your results should resemble Figure 6.9.

**Figure 6.9**
Sorted student database (Department A to Z, GPA largest to smallest).

| | A | B | C | D | E | F | G |
|---|---|---|---|---|---|---|---|
| 1 | | | | | | | |
| 2 | | Last Name | First Name | GPA | Department | Class | |
| 3 | | Ault | Linda | 3.78 | Chemical | Senior | |
| 4 | | Carter | Frank | 3.12 | Chemical | Senior | |
| 5 | | Noland | Rudy | 2.98 | Chemical | Senior | |
| 6 | | Kirk | William | 3.35 | Civil | Sophomore | |
| 7 | | Carter | Christine | 2.78 | Civil | Junior | |
| 8 | | Murray | Susie | 3.92 | Electrical | Senior | |
| 9 | | Phillips | Casey | 3.51 | Electrical | Junior | |
| 10 | | Garcia | Rob | 3.21 | Electrical | Sophomore | |
| 11 | | Hahn | Eric | 3.87 | Mechanical | Junior | |
| 12 | | Madison | Richard | 3.41 | Mechanical | Junior | |
| 13 | | | | | | | |

## 6.5 FILTERING DATA

Excel has several mechanisms for locating records that match specified *criteria*. For example, you may be interested in reviewing the students with Chemical Engineering majors. After you select **Chemical** as the desired criterion, Excel displays only those records with Department field equal to "Chemical."

The process of limiting the visible records on the basis of some criteria is called *filtering*. There are three methods for filtering a database in Excel:

- AutoFilter
- Custom AutoFilter
- Advanced Filter

Filtering does not delete records from a database; it only hides records that do not meet the filter criteria.

### 6.5.1 Using the AutoFilter

The *AutoFilter* (or *Filter*)[2] allows you to filter records while viewing the database on the worksheet. It is very quick and intuitive.

To activate the AutoFilter feature, follow these steps:

1. Select a data cell within the database.
2. Use any of these methods:

   a. From the Ribbon's Home tab: **Home tab → Editing group → Sort & Filter drop-down menu → Filter button.**
   b. From the Ribbon's Data tab: **Data tab → Sort & Filter group → Filter button.**
   c. Excel 2003: Data → Filter → AutoFilter.

---

[2] The term *AutoFilter* was used extensively in Excel 2003, and still appears in some dialog boxes in Excel 2010. The simpler term *Filter* is more common in Excel 2007 and 2010.

A small arrow will appear in the heading of each column (see Figure 6.10). When you click on one of the arrows, a small drop-down menu will appear that contains the possible choices for that field. Figure 6.10 depicts the student database with the AutoFilter option turned on. The drop-down menu for the Department field is shown, with the filter criterion set for **Civil**.

**Figure 6.10**
The AutoFilter, preparing to filter on Department = "Civil".

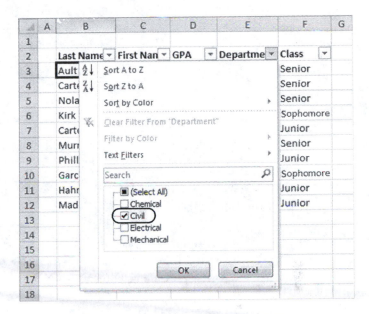

When the **OK** button is clicked, only the records with Department equal to **Civil** will be displayed (see Figure 6.11). The small arrow at the head of the Department column now shows a small funnel icon to signify that this column is filtering some records.

**Figure 6.11**
The filtered database showing only records where Department = "Civil".

To remove the filter and display all records, use one of these methods:

- From the Ribbon's Home tab: **Home tab → Editing group → Sort & Filter drop-down menu → Clear button**.
- From the Ribbon's Data tab: **Data tab → Sort & Filter group → Clear button**.
- Excel 2003: Data → Filter → Show All.

**PRACTICE!**

Practice using the AutoFilter. Use the AutoFilter to select all seniors in Electrical Engineering. Your result should look like Figure 6.12.

**Figure 6.12**
Filtering for Electrical Engineering majors.

| | A | B | C | D | E | F | G |
|---|---|---|---|---|---|---|---|
| 1 | | | | | | | |
| 2 | | Last Name ▼ | First Nam ▼ | GPA ▼ | Departme ▼ | Class ▼ | |
| 8 | | Murray | Susie | 3.92 | Electrical | Senior | |
| 9 | | Phillips | Casey | 3.51 | Electrical | Junior | |
| 10 | | Garcia | Rob | 3.21 | Electrical | Sophomore | |
| 13 | | | | | | | |

When you are done filtering, use any of these approaches to deactivate (toggle) the AutoFilter.

- From the Ribbon's Home tab: **Home tab → Editing group → Sort & Filter drop-down menu → Filter button**.
- From the Ribbon's Data tab: **Data tab → Sort & Filter group → Filter button**.
- Excel 2003: Data → Filter → AutoFilter.

### 6.5.2 Using the Custom AutoFilter

The AutoFilter options that you have learned so far are fine if you want to make an exact match. However, in many cases, you will want to specify a range. For example, you may want to view the students who have GPAs greater than 3.5. One way to specify ranges is to use the *Custom AutoFilter* dialog box. To use the Custom AutoFilter option to specify students with a GPA > 3.5, follow this procedure:

1. Select a data cell within the database.
2. Activate the AutoFilter using any of the methods listed above.
3. Open the AutoFilter menu for the GPA column (using the arrow icon in the column heading).
4. Select **Number Filters . . .** from the GPA drop-down menu (see Figure 6.13).
5. Select **Custom Filter . . .** as shown in Figure 6.13. The Custom AutoFilter dialog box will open, as shown in Figure 6.14.
6. Complete the Custom AutoFilter dialog box data entry fields, as shown in Figure 6.14, to filter for GPAs > 3.5:
   a. Left drop-down list, choose **is greater than**
   b. Right text field, enter **3.5**
7. Click **OK** to close the Custom AutoFilter dialog box and filter the data.

The result of applying the custom filter is shown in Figure 6.15.

**Figure 6.13**
Requesting a custom filter on GPA.

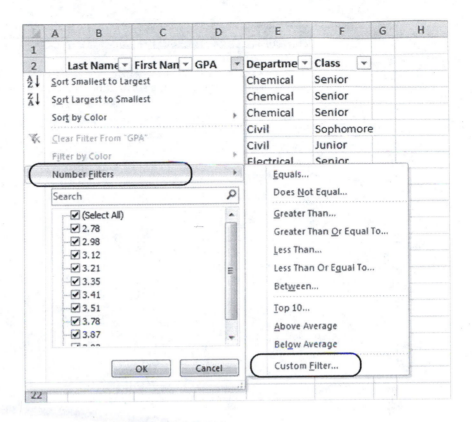

**Figure 6.14**
Custom AutoFilter dialog box.

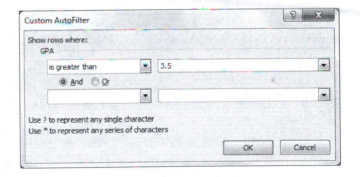

**Figure 6.15**
The database filtered for GPAs > 3.5.

| | A | B | C | D | E | F | G |
|---|---|---|---|---|---|---|---|
| 1 | | | | | | | |
| 2 | | Last Name ▼ | First Nam ▼ | GPA ⏷ | Departme ▼ | Class ▼ | |
| 3 | | Ault | Linda | 3.78 | Chemical | Senior | |
| 8 | | Murray | Susie | 3.92 | Electrical | Senior | |
| 9 | | Phillips | Casey | 3.51 | Electrical | Junior | |
| 11 | | Hahn | Eric | 3.87 | Mechanical | Junior | |
| 13 | | | | | | | |

## PRACTICE!

Practice using the Custom AutoFilter function. You can use it to specify a simple logical expression. The Custom AutoFilter dialog box allows you to join two logical conditions with an *And operator* or an *Or operator*.

If you choose **And**, then both conditions must be true for the record to be displayed. If you choose **Or**, then the record will be displayed if either of the conditions is met.

Use a Custom AutoFilter to select all students with GPAs between 3.0 and 3.5. The Custom AutoFilter dialog box for this filter is illustrated in Figure 6.16. The filtered result is shown in Figure 6.17.

**Figure 6.16**

Creating a custom filter to find students with GPAs between 3.0 and 3.5.

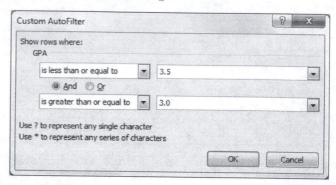

| ◢ | A | B | C | D | E | F | G |
|---|---|---|---|---|---|---|---|
| 1 | | | | | | | |
| 2 | | Last Name ▾ | First Nan ▾ | GPA ▾ | Departme ▾ | Class ▾ | |
| 4 | | Carter | Frank | 3.12 | Chemical | Senior | |
| 6 | | Kirk | William | 3.35 | Civil | Sophomore | |
| 10 | | Garcia | Rob | 3.21 | Electrical | Sophomore | |
| 12 | | Madison | Richard | 3.41 | Mechanical | Junior | |
| 13 | | | | | | | |

**Figure 6.17**

The filtered database: students with GPA between 3.0 and 3.5.

Next, clear the GPA filter and create a custom filter to display all students from either Electrical or Chemical Engineering. Your result should look like Figure 6.18.

| ◢ | A | B | C | D | E | F | G |
|---|---|---|---|---|---|---|---|
| 1 | | | | | | | |
| 2 | | Last Name ▾ | First Nan ▾ | GPA ▾ | Departme ▾ | Class ▾ | |
| 3 | | Ault | Linda | 3.78 | Chemical | Senior | |
| 4 | | Carter | Frank | 3.12 | Chemical | Senior | |
| 5 | | Noland | Rudy | 2.98 | Chemical | Senior | |
| 8 | | Murray | Susie | 3.92 | Electrical | Senior | |
| 9 | | Phillips | Casey | 3.51 | Electrical | Junior | |
| 10 | | Garcia | Rob | 3.21 | Electrical | Sophomore | |
| 13 | | | | | | | |

**Figure 6.18**

The filtered database: Chemical or Electrical engineering majors.

To remove the filter and display all records, use one of these methods:

- From the Ribbon's Home tab: **Home tab → Editing group → Sort & Filter drop-down menu → Clear button**.
- From the Ribbon's Data tab: **Data tab → Sort & Filter group → Clear button**.
- Excel 2003: Data → Filter → Show All.

### 6.5.3 Using Wild-Card Characters

A greater range of choices may be made by using the question mark (?) or asterisk (*) characters as wild-card characters. A ***wild-card character*** is used as a placeholder that can be filled by any legitimate character.

The question mark is used to represent the replacement of any single character. For example, the logical expression

$$\text{Department} = \text{????ical}$$

will return all departments with exactly four characters followed by *ical*. In our student database, Chemical would be displayed, but Mechanical, Electrical, and Civil would not, as they do not contain exactly four characters followed by *ical*.

The asterisk is used to represent the replacement of zero or more characters. For example, the logical expression

$$\text{Department} = \text{*ical}$$

will return all departments ending in *ical*, no matter how many letters precede *ical* (including zero letters). In our student database, Chemical, Electrical, and Mechanical would be displayed, but Civil would not.

Some wild-card functions can also be performed by using the following conditions in the Custom AutoFilter dialog:

- Begins with
- Ends with
- Contains
- Does not begin with
- Does not end with
- Does not contain

### PRACTICE!

Practice using the wild-card characters and menu selections in the Custom AutoFilter dialog box. Be sure to clear the filter between each exercise. Here are two methods of selecting all students whose last name begins with an S:

*Method 1.*

**Last Name**    **begins with**    S

*Method 2.*

**Last Name**    **equals**    S*

Here are two methods for selecting all students whose last name ends in an N and does not contain an H:

*Method 1.*

|             |                  |     |
|-------------|------------------|-----|
| **Last Name** | **ends with** | N |
|             | **And**          |     |
| **Last Name** | **does not contain** | H |

*Method 2.*

|             |                  |     |
|-------------|------------------|-----|
| **Last Name** | **equals**    | *N |
|             | **And**          |     |
| **Last Name** | **does not equal** | *H* |

Note that the Custom AutoFilter box is not case sensitive.

Try creating a Custom AutoFilter for selecting all students whose last name ends in an N and does not contain an H. Your result should look like Figure 6.19.

| ⏢ | A | B | C | D | E | F | G |
|----|---|---|---|---|---|---|---|
| 1  |   |           |          |      |            |        |   |
| 2  |   | Last Name▼ | First Nan▼ | GPA ▼ | Departme▼ | Class ▼ |   |
| 12 |   | Madison   | Richard  | 3.41 | Mechanical | Junior |   |
| 13 |   |           |          |      |            |        |   |

**Figure 6.19**
Results after filtering.

## 6.5.4 Using the Advanced Filter

The filtering methods that you have been shown so far allow a great deal of flexibility. By adding filters to multiple fields and by using the Custom AutoFilter, wildcards, and logical operators, you can build relatively complex filters.

However, there are some cases that can't be handled by these methods. Suppose that the engineering departments have different GPA requirements. We want to find students who meet the following criteria:

**Electrical Engineering with GPA > 3.6**
**Or**
**Chemical Engineering with GPA > 3.0**
**Or**
**Civil Engineering with GPA > 3.2**
**Or**
**Mechanical Engineering with GPA > 3.4**

You can't create a filter that solves this request by using the AutoFilter. You will need to use the ***Advanced Filter***, which allows you to build more complex queries.

The term ***query*** can be used as a noun or a verb. As a verb, it means to ask a question. When we query a database, we are asking the database program to return all records that meet our criteria. As a noun, the query is the list of criteria that we want the database program to use to identify applicable records.

To use the Advanced Filter, you must first set up a Criteria table. A ***Criteria table*** is just what it sounds like, a table of criteria that must be met for a filter to occur. To create a Criteria table, perform these steps:

1. Copy the field names from your database to another location in the same work-sheet. (Leave at least one blank row of cells between the Criteria table and the database.)
2. Type in the criteria that must be met for your filter.

### 6.5.5 Logic Within Rows

All criteria within a single row must be met for a match to occur. This is equivalent to a logical *And* operator.

### 6.5.6 Logic Between Rows

A Criteria table may have more than one active row. A match on any row in a Criteria table may be met for a match to occur. This is equivalent to a logical *Or* operator.

As an example, Figure 6.20 shows a Criteria table to select students in each major who meet the department's GPA requirement.

**Figure 6.20**
Criteria Table and Database before filtering.

| | A | B | C | D | E | F | G |
|---|---|---|---|---|---|---|---|
| 1 | | | | | | | |
| 2 | | | Criteria Table | | | | |
| 3 | | Last Name | First Name | GPA | Department | Class | |
| 4 | | | | >3.6 | Electrical | | |
| 5 | | | | >3.0 | Chemical | | |
| 6 | | | | >3.2 | Civil | | |
| 7 | | | | >3.4 | Mechanical | | |
| 8 | | | | | | | |
| 9 | | | Database | | | | |
| 10 | | Last Name | First Name | GPA | Department | Class | |
| 11 | | Ault | Linda | 3.78 | Chemical | Senior | |
| 12 | | Carter | Frank | 3.12 | Chemical | Senior | |
| 13 | | Noland | Rudy | 2.98 | Chemical | Senior | |
| 14 | | Kirk | William | 3.35 | Civil | Sophomore | |
| 15 | | Carter | Christine | 2.78 | Civil | Junior | |
| 16 | | Murray | Susie | 3.92 | Electrical | Senior | |
| 17 | | Phillips | Casey | 3.51 | Electrical | Junior | |
| 18 | | Garcia | Rob | 3.21 | Electrical | Sophomore | |
| 19 | | Hahn | Eric | 3.87 | Mechanical | Junior | |
| 20 | | Madison | Richard | 3.41 | Mechanical | Junior | |
| 21 | | | | | | | |

Now that we have set up the Criteria table, we can use the Advanced Filter dialog box to build a filter by using the Criteria table. To use the Advanced Filter, follow these steps:

1. Click on any data cell in the database to select it. Excel will automatically detect all of the database records.
2. Open the Advanced Filter dialog box with Ribbon options: **Data tab → Sort & Filter group → Advanced button**. (Excel 2003: Data → Filter → Advanced Filter.) The Advanced Filter dialog box is shown in Figure 6.21.

3. Indicate the cell range containing the Criteria table, including the column headings. (Cells B3:F7 in this example.)
4. Select the **Action: Filter the list, in-place**.
5. Click **OK** to close the Advanced Filter dialog box and filter the data.

The results are shown in Figure 6.22.

**Figure 6.21**
The Advanced Filter dialog box.

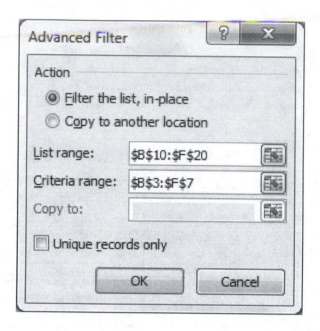

**Figure 6.22**
The database after applying the Advanced Filter.

| | A | B | C | D | E | F | G |
|---|---|---|---|---|---|---|---|
| 1 | | | | | | | |
| 2 | | | | Criteria Table | | | |
| 3 | | Last Name | First Name | GPA | Department | Class | |
| 4 | | | | >3.6 | Electrical | | |
| 5 | | | | >3.0 | Chemical | | |
| 6 | | | | >3.2 | Civil | | |
| 7 | | | | >3.4 | Mechanical | | |
| 8 | | | | | | | |
| 9 | | | | Database | | | |
| 10 | | Last Name | First Name | GPA | Department | Class | |
| 11 | | Ault | Linda | 3.78 | Chemical | Senior | |
| 12 | | Carter | Frank | 3.12 | Chemical | Senior | |
| 14 | | Kirk | William | 3.35 | Civil | Sophomore | |
| 16 | | Murray | Susie | 3.92 | Electrical | Senior | |
| 19 | | Hahn | Eric | 3.87 | Mechanical | Junior | |
| 20 | | Madison | Richard | 3.41 | Mechanical | Junior | |
| 21 | | | | | | | |

## KEY TERMS

| | | |
|---|---|---|
| Advanced Filter | data entry form | filtering |
| *And* operator | database | *Or* operator |
| AutoFilter | database management | query |
| criteria | system (DBMS) | record |
| Criteria table | field | sort |
| Custom AutoFilter | field name | wild-card character (?,*) |

## SUMMARY

### Delimiting a Database in Excel

- Leave a perimeter of blank cells around the database region.
- Explicitly name the region.

### Recommendations for Using Databases within Excel

- Maintain only one database per worksheet.
- Use unique column headings.
- Leave an empty column to the right of the database and an empty row at the bottom of the database.
- Do not use cells to the right or left of the database for other purposes.

### Sorting a Database

1. Select any data cell in the database.
2. Open the Sort dialog box: **Data tab** ➜ **Sort & Filter group** ➜ **Sort button**. The Sort dialog box will open.
3. Select the field name and sort order.
4. Click **Add Level** if more than one field is included in the sort, then select the next field name and sort order.
5. If your database includes column headings, be sure that the **My data has headers** box is checked.
6. Click **OK**.

### Filtering a Database

There are three methods for filtering a database in Excel:

- AutoFilter
- Custom AutoFilter
- Advanced Filter

### Activating and Deactivating the AutoFilter (toggles)

1. Select a data cell within the database.
2. **Data tab** ➜ **Sort & Filter group** ➜ **Filter button**.

**Filtering with the AutoFilter**

1. Click on the small arrow on the desired column heading.

2. Select the desired filter criterion from the drop-down menu.

**Clearing the AutoFilter**

* **Data tab → Sort & Filter group → Clear button**

**Filtering with the Custom AutoFilter**

1. Select a data cell within the database.

2. Activate the AutoFilter.

3. Open any AutoFilter menu (using the arrow icon in the column heading).

4. Select **Number Filters . . .** or **Text Filters . . .** from the drop-down menu.

5. Select **Custom Filter . . .** The Custom AutoFilter dialog box will open.

6. Complete the Custom AutoFilter dialog box.

7. Click **OK.**

**Wild-Card Characters**

* ? – replacement of any single character.
* * – replacement of zero or more characters.

**Filtering with the Advanced Filter**

1. Create a Criteria table.
   a. Logic within rows is equivalent to a logical *And* operator.
   b. Logic between rows is equivalent to a logical *Or* operator.

2. Click on any data cell in the database to select it.

3. Open the Advanced Filter dialog box: **Data tab → Sort & Filter group → Advanced button.**

4. Indicate the cell range containing the Criteria table, including the column headings.

5. Select the **Action: Filter the list, in-place** is the more commonly used.

6. Click **OK.**

## PROBLEMS

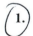 1. Figure 6.23 shows the thermal properties of sundry materials. The construction column indicates whether or not the material is used in typical residential construction. Type this data into a worksheet, or download the worksheet from the author's website:

   http://www.chbe.montana.edu/IntroExcel

From the web page, select the item labeled *Thermal Properties*.

Use this worksheet to solve the following:

a. Sort the worksheet by **Specific Heat** in ascending order. Where do the blank entries get placed after the sort?

b. Sort the worksheet by the **Construction** field with the **yes** category at the top. Within each Construction category, sort in descending order of **Density**.

**Figure 6.23**
Thermal properties of various materials.

| | A | B | C | D | E | F |
|---|---|---|---|---|---|---|
| 1 | Thermal Properties of Various materials (at 20°C) | | | | | |
| 2 | | | | | | |
| 3 | Material | Construction | Density - $\rho$ (kg/m$^3$) | Specific Heat - $c_p$ (J/kg.K) | Conductivity - k (J/s.m.°C) | |
| 4 | Aluminum | yes | 2700 | 896 | 237.00 | |
| 5 | Bronze | no | 8670 | 343 | 26.00 | |
| 6 | Concrete | yes | 500 | 840 | 0.13 | |
| 7 | Copper | no | 8930 | 383 | 400.00 | |
| 8 | Glass | yes | 2800 | 800 | 0.81 | |
| 9 | Ice | no | 910 | 57 | 202.00 | |
| 10 | Plaster | yes | 1800 | 112 | 0.81 | |
| 11 | Polystyrene | yes | 1210 | | 0.04 | |
| 12 | Wood (pine) | yes | 420 | 2700 | 0.15 | |
| 13 | Wool Insulation | yes | 200 | | 0.04 | |
| 14 | | | | | | |

   **c.** Use the AutoFilter to show only **Construction** materials.

   **d.** Use the Custom AutoFilter to show only construction materials with a **Density** < 1000 and **Conductivity** < 1.00.

   **e.** Use the Advanced Filter to show **Construction** materials with a **Specific Heat** > 800 and non-construction materials (**Construction** = no) with a **Specific Heat** > 300. Show your criteria table.

   **f.** Can you solve the last problem using only the AutoFilter, not the Advanced Filter?

Describe the actions you performed to solve each part of this problem.

**2.** A common database used by many people is an address book. It contains names, addresses, phone numbers, and sometimes other information such as birth dates and anniversaries. But paper-based address books can be hard to keep in alphabetical order as names are added and deleted, and sometimes they don't have enough fields for all of the phone numbers and email addresses we use today.

   Create an address book database in Excel containing the fields listed below. You can add other fields if you like to keep track of additional information.

- Last Name
- First Name
- Street Address
- City
- State
- Postal Code
- Email
- Home Phone
- Work Phone
- Cell Phone

Add at least three people to the address book, and sort the data by last name.

**3.** When students need to work together on projects, conflicts can arise because some students like to wait until the last possible minute, some like to get the projects done as soon as possible, and some like to work at a slow and steady pace. Also, scheduling meetings can be a problem because some students like to get together in the afternoons to work on projects, some like evenings, and others want to do the work only on the weekends.

Enter the database shown in Figure 6.24, and use Excel's database filtering to identify the following:

a. Groups of students who all like to get projects "done and over with" as soon as possible.

b. Groups of students who all like to work on group projects in the afternoons.

**Figure 6.24**

Student work preferences database.

|  | A | B | C | D |
|---|---|---|---|---|
| 1 | Student Preferences Database | | | |
| 2 | | | | |
| 3 | Name | Preferred Work Style | Preferred Work Time | |
| 4 | Seth | Done and Over With | Evenings | |
| 5 | Alicia | Slow and Steady | Afternoons | |
| 6 | Allyson | Last Minute | Evenings | |
| 7 | Terry | Last Minute | Afternoons | |
| 8 | Sam | Done and Over With | Weekends | |
| 9 | Brittany | Last Minute | Weekends | |
| 10 | Treair | Slow and Steady | Evenings | |
| 11 | Rob | Last Minute | Afternoons | |
| 12 | | | | |

4. Most data tables can be treated as databases and sorted or filtered. For example, consider the properties of liquid water tabulated in Figure 6.25. Create the data table shown in Figure 6.25, then perform these tasks:

a. Filter the database to show data for temperatures greater than or equal to 10°C and less than 20°C.

b. Filter the database to show data with thermal conductivities greater than 0.6 W/m K.

**Figure 6.25**

Properties of liquid water (approximate values).

|  | A | B | C | D | E |
|---|---|---|---|---|---|
| 1 | Properties of Liquid Water | | | | |
| 2 | | | | | |
| 3 | T (°C) | Visc. (cP) | SG | Th. Cond. (W/m K) | |
| 4 | 0 | 1.8 | 1.000 | 0.554 | |
| 5 | 5 | 1.5 | 1.000 | 0.565 | |
| 6 | 10 | 1.3 | 1.000 | 0.577 | |
| 7 | 15 | 1.1 | 0.999 | 0.587 | |
| 8 | 20 | 1.0 | 0.998 | 0.597 | |
| 9 | 25 | 0.9 | 0.997 | 0.606 | |
| 10 | 30 | 0.8 | 0.996 | 0.615 | |
| 11 | 35 | 0.7 | 0.994 | 0.623 | |
| 12 | 40 | 0.6 | 0.992 | 0.630 | |
| 13 | | | | | |

5.  One problem with having lots of books is keeping track of them. The database shown in Figure 6.26 is one way to keep records.

**Figure 6.26**
Personal library management database.

| | A | B | C | D | E |
|---|---|---|---|---|---|
| 1 | **Book** | **Type** | **Status** | **Loaned To** | |
| 2 | Popular Myths about Engineers | Humor | In Library | | |
| 3 | 10,000 Engineer Jokes | Humor | On Loan | Paula | |
| 4 | Statics and Dynamics for Dummies | Text | In Library | | |
| 5 | Complete and Concise Steam Tables | Reference | On Loan | Max | |
| 6 | Properties of Liquid Water | Reference | In Library | | |
| 7 | | | | | |

List the steps you would use in Excel to filter the database to display only reference books that are out on loan.

6.  Figure 6.27 shows the initial stages of a database for holding a grocery list. If this list were maintained, any of your roommates could easily filter the database to show only needed items and quickly print a grocery list.

**Figure 6.27**
Grocery list database.

| | A | B | C | D | E | F |
|---|---|---|---|---|---|---|
| 1 | **Grocery List** | | | | | |
| 2 | | | | | | |
| 3 | **Item** | **Category** | **Size** | **On Hand** | **Status** | |
| 4 | Bread | Bakery | Loaf | 0 | Needed | |
| 5 | Lunch Meat, Turkey | Deli | 1 pound, sliced | 2 | Not Needed | |
| 6 | Lunch Meat, Beef | Deli | 1 pound, sliced | 1 | Not Needed | |
| 7 | Mayo | Condiments | Jar, 8 oz. | 0 | Needed | |
| 8 | | | | | | |

Create your own grocery list database and practice filtering, as follows:

a. Filter for **Needed** items.

b. Filter for items that are running low by filtering for **On Hand** $< 2$.

# Collaborating with Other Engineers

## Section

## Objectives

*After reading this chapter, you should be able to perform the following tasks:*

- Track revisions in an Excel document.
- Share workbooks among team members.
- Insert comments in an Excel document.
- Transfer worksheet data to and from other applications.
- Use a password to restrict ability to open a file.
- Use a password to restrict ability to write to a file.
- Use a password to restrict access to a worksheet.

## 7.1 THE COLLABORATIVE DESIGN PROCESS

Engineering design is the process of devising an effective and efficient solution to a problem. The solution may take the form of a component, a system, or a process. Engineers generally solve problems by collaborating with others as members of a team. As a student, you will undoubtedly be asked to participate in collaborative projects with other students.

You may or may not have experience working on a team. If a team works together effectively, then more can be accomplished by the team than through any individual effort (or even the sum of individual efforts). If team members do not work together effectively, however, the group can become mired in power struggles and dissension. When this occurs, one of two things usually happens. Either the team makes little progress toward its goals, or a small subgroup of the team takes charge and does all of the work. Some guidelines for being an effective team member are presented at the end of this chapter, in the "Professional Success" section.

### 7.1.1 Microsoft Excel and Collaboration

The ability to work well on a team can best be learned by participating in a successful team effort. Microsoft Excel includes several tools that can help to solve one of the most burdensome technical tasks of group collaboration—the preparation of the team workbook. In the past, collaborative workbook preparation has been extremely difficult. The result has been that the task is usually assigned to one or two team members. New features of Microsoft Excel make it feasible for the whole team to participate in the composition and revision of a workbook. The Ribbon includes a *Review tab* that is all about working with others. Learning to use these features will require some time and practice on your team's part. The rewards, however, will be well worth the effort.

## 7.2 TRACKING CHANGES

One problem that arises in workbook preparation by a team is keeping track of revisions. For example, one team member may be given the task of revising a portion of the team project. After the revisions are made, the team will meet and approve some, or all, of the revisions. Then one of the team members will incorporate the accepted changes into the workbook.

Excel has a feature called ***Tracking Changes*** that will not only mark revisions, but will also keep track of who is making each revision. The worksheet may be printed showing both the original text and the new revisions. Revisions may then be globally accepted or selectively accepted into the workbook.

### 7.2.1 Highlighting Changes

To activate the tracking changes feature, follow these steps:

1. Open the Highlight Changes dialog box (shown in Figure 7.1) using Ribbon options: **Review tab → Changes group → Track Changes drop-down menu → Highlight Changes . . . button**. (Excel 2003: Tools → Track Changes → Highlight Changes.)
2. Check the **Track Changes while editing** box. This will make the workbook available to others (*shared*), and it will turn on ***history tracking***. You will now have access to the next three boxes and drop-down lists, which allow you to limit the changes that are highlighted by time, user, and worksheet region.
3. Check the **When** box and select **All** from the drop-down list. All subsequent changes to the workbook will be tracked.
4. Check the **Who** box and select **Everyone** from the drop-down list to track changes by every user of the worksheet.
5. Check the **Where** box and use the **Jump to Worksheet** button to select the entire worksheet. (Select the entire worksheet by clicking the corner of the cell headings, between the A and 1 headings as illustrated in Figure 7.2.)

6. Check the **Highlight changes on screen** box so that all changes to the worksheet will be easy to see.

Note: The alternative is to list changes on a new sheet. This can be useful when there are extensive changes.

7. Click **OK** to close the Highlight Changes dialog box.

**Figure 7.1**
The Highlight Changes dialog box.

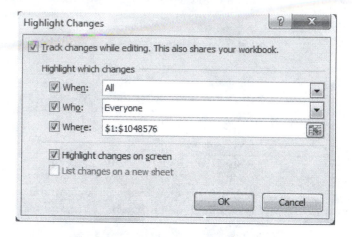

**Figure 7.2**
Selecting the entire worksheet.

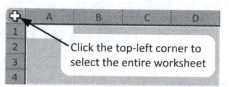

Click the top-left corner to select the entire worksheet

Make a change on your worksheet and note what happens. Any modified cells are outlined in blue, and a small blue tab is placed in the upper left corner of the cell. These are called *revision marks*. (See Figure 7.3.)

**Figure 7.3**
Revision mark in cell B10.

| | A | B | C | D | E | F | G | H |
|---|---|---|---|---|---|---|---|---|
| 1 | Body Mass Index (BMI) Calculations | | | | | | | |
| 2 | | | | | | | | |
| 3 | Height (in) | Weight (lb) | BMI | Status | | Status Codes | | |
| 4 | 62 | 136 | 24.9 | Normal | | 0.0 | Underweight | |
| 5 | 74 | 195 | 25.0 | Overweight | | 18.5 | Normal | |
| 6 | 68 | 155 | 23.6 | Normal | | 25.0 | Overweight | |
| 7 | 71 | 130 | 18.1 | Underweight | | 30.0 | Obese | |
| 8 | 65 | 160 | 26.6 | Overweight | | | | |
| 9 | 58 | 114 | 23.8 | Normal | | | | |
| 10 | 76 | 195 | | | | | | |
| 11 | | | | | | | | |

Revision mark

### 7.2.2 Creating an Identity

As you review a document that other team members have revised, you can see who made the revision, the date and time of the revision, and the previous contents of the cell. The identity feature works only if each reviewer has provided an identity to the Excel application. To identify yourself to Excel, do the following:

1. Open the Excel Options dialog box shown in Figure 7.4.
   - Excel 2010: **File tab ➔ Options button**
   - Excel 2007: **Office button ➔ Excel Options button**
   - Excel 2003: Tools ➔ Options
2. Use the **General** panel, and the **Personalize your copy of Microsoft Office** section. (Excel 2003: General tab.)
3. Enter your name in the **User name** box.
4. Click **OK** to close the Excel Options dialog box.

**Figure 7.4**

Identifying yourself using the Excel Options dialog box.

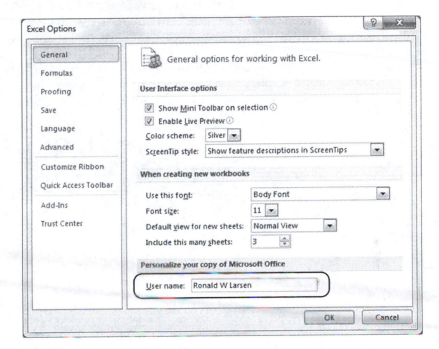

Your identity will now be attached to any revisions that you make to a worksheet. You can test the feature by making a change. If you move the mouse over the revision mark, a comment is displayed (shown in Figure 7.5) that indicates who made the change, and when.

**Figure 7.5**

Example of a revision mark with information about who made the change, and when.

### 7.2.3 Incorporating or Rejecting Revisions

Revision marks are not wholly incorporated into the document until they are reviewed and then accepted or rejected. Revision marks can be reviewed by using the Accept or Reject Changes dialog box. To accept or reject changes, follow these steps:

1. Open the Select Changes to Accept or Reject dialog box (shown in Figure 7.6) using Ribbon options: **Review tab → Changes group → Track Changes drop-down menu → Accept/Reject Changes . . . button**. (Excel 2003: Tools → Track Changes → Accept or Reject Changes.)
2. Check the **When** box and select **Not Yet Reviewed** from the drop-down list. This will allow you to review all new (not-yet-reviewed) changes.
3. Check the **Who** box and select **Everyone** from the drop-down list to see the changes made by all worksheet users.
4. Check the **Where** box and select the entire worksheet.
5. Click **OK** to close the Select Changes to Accept or Reject dialog box.

**Figure 7.6**
Select Changes to Accept or Reject dialog box.

Excel will now guide you through each selected revision and give you the opportunity to accept or reject the revision. The Accept or Reject Changes dialog box will appear, as shown in Figure 7.7. You will be guided through the revisions one at a time, unless you select **Accept All** or **Reject All**.

**Figure 7.7**
The Accept or Reject Changes dialog box.

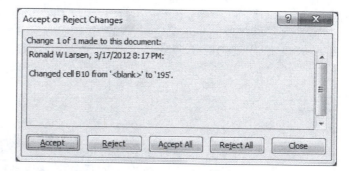

## 7.3 ADDING COMMENTS TO A DOCUMENT

At times, a reviewer may want to attach notes or *comments* to a cell without changing the contents of the cell. To add a comment to a cell, follow this procedure:

1. Select the cell that you want to comment on.
2. Click the **New Comment** button on the Review tab: **Review tab → Comments group → New Comment button**. (Excel 2003: Insert → Comment.)
3. Type in the text of the comment, then click outside the comment box when finished.

A cell that is attached to a comment will be marked with a small red tab in the upper right-hand corner of the cell (see cell B3 in Figure 7.8). When the mouse cursor is moved over the red tab, the comment will be displayed.

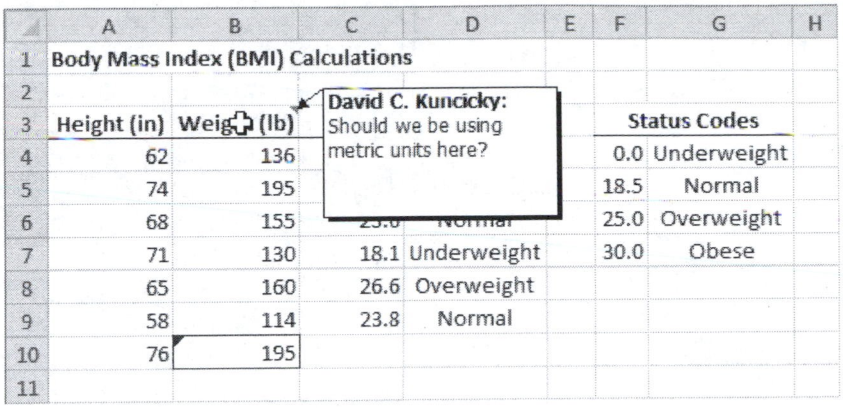

**Figure 7.8**
Example of a comment added to cell B3.

| | A | B | C | D | E | F | G | H |
|---|---|---|---|---|---|---|---|---|
| 1 | Body Mass Index (BMI) Calculations | | | | | | | |
| 2 | | | | | | | | |
| 3 | Height (in) | Weight (lb) | David C. Kuncicky: Should we be using metric units here? | | | Status Codes | | |
| 4 | 62 | 136 | | | | 0.0 | Underweight | |
| 5 | 74 | 195 | | | | 18.5 | Normal | |
| 6 | 68 | 155 | 23.6 | Normal | | 25.0 | Overweight | |
| 7 | 71 | 130 | 18.1 | Underweight | | 30.0 | Obese | |
| 8 | 65 | 160 | 26.6 | Overweight | | | | |
| 9 | 58 | 114 | 23.8 | Normal | | | | |
| 10 | 76 | 195 | | | | | | |
| 11 | | | | | | | | |

Displayed comments may be viewed, edited, or deleted. You can view comments one at a time by moving the mouse cursor over each comment tab (Figure 7.8), or you can also toggle the **Show All Comments** button on the Ribbon's Review tab to see all comments on a worksheet. (Excel 2003: **Show/Hide All Comments** button on the Reviewing toolbar.)

To edit or delete a displayed comment from the Ribbon:

1. Click on the comment to select it, then
2. Use the **Edit Comment** or **Delete Comment** buttons on the Review tab.

In Excel 2003, use these steps (these steps work in Excel 2007 and 2010, too):

1. Right-click on the cell containing the comment.
2. Select **Edit Comment** or **Delete Comment** options from the pop-up menu.

## 7.4 MAINTAINING SHARED WORKBOOKS

Excel provides a mechanism that allows several users to simultaneously *share* a workbook over a network. A ***shared workbook*** must reside in a shared folder on the network. Other access restrictions may apply, depending on your local network setup.

Once you are able to share a workbook, different users can view and modify the workbook at the same time. Sharing a workbook clearly requires some protocol among the group in order to keep several users from overwriting each other's work. Sharing a workbook is most effective if simultaneous users edit different parts of the workbook. Excel can be set up to keep a history of changes to a shared workbook, and previous versions may be recalled if necessary.

### 7.4.1 Sharing a Workbook

Follow these steps to share a workbook:

1. Click the Share Workbook button on the Review tab to open the Share Workbook dialog box (Figure 7.9): **Review tab ➔ Changes group ➔ Share Workbook button**. (Excel 2003: Tools ➔ Share Workbook.)
2. Select the Editing tab.
3. Check the box labeled **Allow changes by more than one user at the same time**.

Once the workbook is shared, this tab can be used to see who is currently editing the workbook.

**Figure 7.9**
The Share Workbook
dialog box.

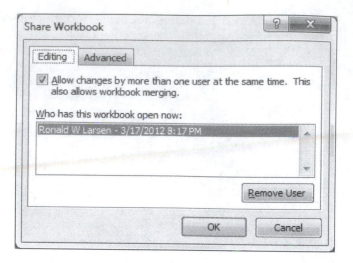

### 7.4.2 Keeping a Change History

Excel can keep a log of changes made by each user of a shared workbook. The log of changes is called a *change history*. To set the options for a change history, follow this procedure:

1. Click the Share Workbook button on the Review tab to open the Share Workbook dialog box (Figure 7.9): **Review tab → Changes group → Share Workbook button.** (Excel 2003: Tools → Share Workbook.)
2. Select the **Advanced tab**, as shown in Figure 7.10.
3. In the section labeled **Track Changes**, set the length of time to keep a change history.

**Figure 7.10**
The Advanced tab on
the Share Workbook
dialog box.

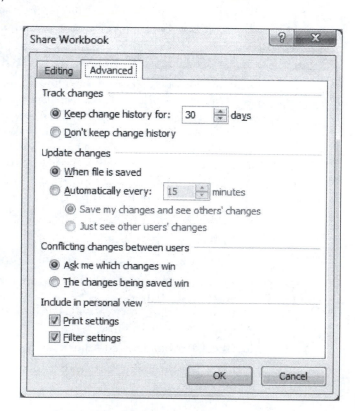

If you decide not to keep a history of changes, select the **Track changes** option labeled **Don't keep change history**.

One reason to turn off the change history or to keep the time duration low is to limit the size of the workbook. A change history can significantly increase the disk space required to store a workbook. There is a trade-off between caution and storage requirements. The use of the change-history feature is not a substitute for regularly backing up a workbook to some other medium, such as a removable disk or tape.

### 7.4.3 Managing Conflicts

If you are about to save a workbook, some of your changes may conflict with pending changes from another user. The **Conflicting changes between users** section of the Advanced tab allows you to specify how you want to resolve conflicts, if at all. If you choose the first option, **Ask me which changes win**, then the Resolve Conflicts dialog box will appear when you attempt to save the file.

You will be prompted to resolve each conflict. If you don't want to resolve conflicts when you save a shared workbook, then select the option labeled **The changes being saved win**. The last user to save conflicting changes wins.

### 7.4.4 Personal Views

The **Include in personal view** section of the Advanced tab allows you to use personal printer or filter settings. When the workbook is saved, a separate personal view is saved for each user.

### 7.4.5 Merging Workbooks

Group members do not always have access to the same network. One scenario that occurs when groups collaborate on a workbook is that each member takes a copy of the workbook home for the evening. Each group member works separately on the workbook, and later the workbooks are merged back into a single document.

Copies of a workbook can be revised and merged only if a change history is being maintained. Be sure to set a sufficient length of time for the change history so that the history doesn't expire before the workbook copies are merged. The number of days is set in the Share Workbook dialog box as shown in Figure 7.10.

Before you can merge workbooks in Excel 2007 and 2010, you need to put the Compare and Merge Workbooks button on the Quick Access Toolbar with these steps (not needed in Excel 2003):

Click the **Customize Quick Access Toolbar** button (small down arrow) at the right side of the Quick Access Toolbar and select **More Commands . . .** from the menu. The Excel Options dialog box will open.

1. Be certain that the **Quick Access Toolbar** panel is open as shown in Figure 7.11 (**Customize** panel in Excel 2007).
2. Select **Commands Not in the Ribbon** from the **Choose Commands from** drop-down list.
3. Select **Compare and Merge Workbooks** from the left selection list.
4. Click the **Add** button.
5. Click **OK** to close the Excel Options dialog box.

Once the **Compare and Merge Workbooks** button is available, follow these steps to merge several copies of a workbook:

1. Open the first copy and then click the **Compare and Merge Workbooks** button on the Quick Access Toolbar. (Excel 2003: Tools ➜ Compare and Merge Workbooks.) You will be prompted to choose a file to merge.
2. Continue to merge files until all copies have been merged into one workbook.

**Figure 7.11**
Adding the Compare and Merge Workbooks button to the Quick Access toolbar.

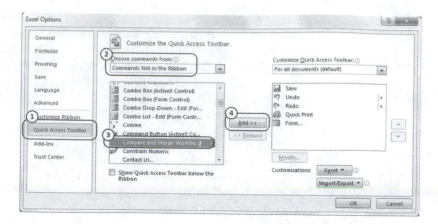

After the workbooks have been merged, you will need to view the change history and selectively choose the changes that you want in the merged workbook.

### 7.4.6 Restrictions for Shared Workbooks

Some features of Excel cannot be used while a workbook is being shared. However, all features can be used if the workbook has sharing turned off. The disadvantage of turning off sharing is that the change history is deleted. When a workbook is shared, a good rule of thumb is to use sound judgment, as not all worksheets or workbooks are meant to be shared. Here's a list of some of the features of Excel that cannot be used when sharing is in effect:

- Creation, modification, or deletion of passwords. Passwords should be set up before the workbook is shared.
- Deletion of worksheets.
- Insertion or modification of charts, pictures, or hyperlinks.
- Insertion or deletion of regions of cells. (Single rows or columns can be deleted.)
- Creation of data tables or pivot tables.
- Insertion of automatic subtotals.

## 7.5 PASSWORD PROTECTION FOR WORKBOOKS

Several levels of *protection* exist for workbooks. Your personal file space may be protected by the network operating system. The folder in which your workbook resides may be protected. These methods are outside the scope of this section. The methods that are discussed here apply only to a single workbook or parts of a workbook.

The methods that follow, in and of themselves, will not prevent another user from copying your workbook; but one of the methods (open access) can prevent another user from viewing the copied workbook.

One way to limit access to a shared workbook is with *password protection*. A variety of password types will be discussed next. In every case, be sure to write down or memorize your password. If you lose a password, you will not be able to retrieve your work.

### 7.5.1 Open Protection

A password can be set that restricts a user from opening a file. This means that an unauthorized user cannot read or print the file by using Excel. This type of access is called *open access*, since it protects a file from being opened. A user will still be able to copy the file. The password protection for open access will also apply to the copied file.

**Note:** You cannot restrict open access on a shared workbook.

To set a password for open access, follow this procedure:

1. Open the Save As dialog:
   * Excel 2010: **File tab → Save As**
   * Excel 2007: **Office button → Save As sub-menu → Excel Workbook option**
   * Excel 2003: File → Save As
2. Choose **Tools → General Options . . .** from the Save As dialog box as shown in Figure 7.12. The General Options dialog box will open, as shown in Figure 7.13.
3. Choose the **Always create backup** option. This option specifies that Excel should create a backup copy of your workbook every time it is saved. Unless you are extremely short of disk space, this is an excellent option!
4. To restrict others from opening your workbook file, type a password in the box labeled **Password to open,** as illustrated in Figure 7.13.
5. You will be prompted to confirm the password by entering it a second time (Figure 7.14). Note that Excel uses case-sensitive passwords.

**Hint:** One of the most common reasons that a password seems to suddenly stop working is that you have the caps lock key turned on.

**Figure 7.12**
Selecting **General Options** . . . from the Tools menu on the Save As dialog.

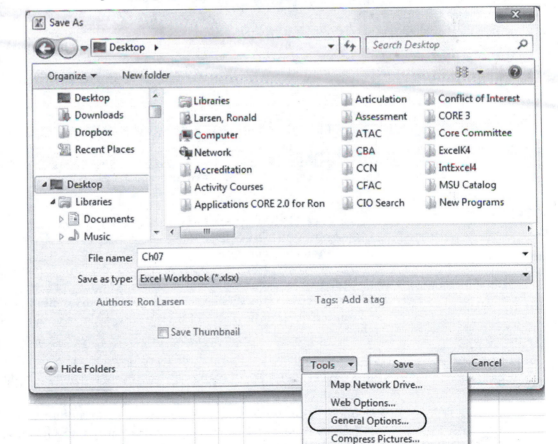

**Figure 7.13**
The General Options
dialog box.

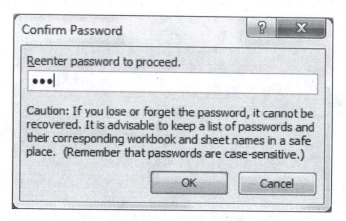

**Figure 7.14**
The Confirm Password
dialog box.

## 7.5.2 Modify or Write Protection

There may be times when you want to allow others to read your workbook (read access), but you do not want anyone to be able to save changes to the workbook. This type of protection is called *modify access* or *write access* because others are not allowed to write the changed file to a storage device. The procedure to set a modify password is very similar. Follow these steps:

1. Open the Save As dialog:
   a. Excel 2010: **File tab → Save As**
   b. Excel 2007: **Office button → Save As sub-menu → Excel Workbook option**
   c. Excel 2003: File → Save As
2. Choose **Tools → General Options . . .** from the Save As dialog box. The General Options dialog box will open, as shown in Figure 7.13.
3. Type a password in the box labeled **Password to modify** (see Figure 7.13).
4. You will be prompted to confirm the password by entering it a second time.

The next time you attempt to open the file, the Password dialog box will appear, as depicted in Figure 7.15. You will be prompted for a password if you want to open the file for write access. A password is not needed to open the file for reading only.

**Note:** A user can open a write-protected file as a read-only file and then save it under a different name. The new file can be modified by the user without a password. So, modify access protects your workbook file from changes, but does not protect you from having others use your work.

**Figure 7.15**
The Password dialog is used to access a password protected workbook.

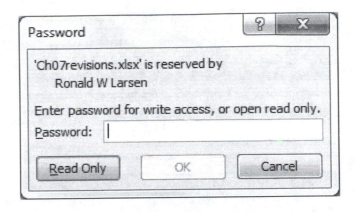

### 7.5.3 Sheet Protection

Protection can be finely tuned. Once you have completed part of a worksheet, you may want to protect it merely to prevent yourself from inadvertently modifying that section. One example of the use of *sheet protection* is to lock cells that contain formulas, while allowing cells that contain input data to be modified.

By default, all cells are locked (but locking cells has no effect until the worksheet is password protected). Before activating worksheet protection, you must unlock the cells that you want to be available after the worksheet is protected. The general procedure for locking down a worksheet is:

1. Unlock the cells that are to be available after the worksheet is locked.
2. Protect the worksheet.

For example, Figure 7.16 shows a worksheet that can be used to solve quadratic equations. The formulas for the two solutions are in cells C9 and C10. The user inputs the three coefficients a, b, and c in cells C4, C5, and C6, respectively. We need to unlock the cells containing the coefficients so that they can still be used after locking the worksheet.

**Note:** The cells holding the coefficient values have been named AA, BB, and CC to make the formulas easier to read. CC was used rather than C because C is a reserved variable name in Excel; it stands for "column."

**Figure 7.16**
Worksheet for solving quadratic equations.

| C9 | ▼ | $f_x$ | =(-BB+SQRT(BB^2-4*AA*CC))/(2*AA) | | | | |
|---|---|---|---|---|---|---|---|
| | A | B | C | D | E | F | G |
| 1 | Solving Quadratic Equations | | | | | | |
| 2 | | | | | | | |
| 3 | | Coefficients | | | | | |
| 4 | | A: | 2 | | multiplies $x^2$ | | |
| 5 | | B: | -4 | | multiplies x | | |
| 6 | | C: | 1 | | constant | | |
| 7 | | | | | | | |
| 8 | | Solutions | | | | | |
| 9 | | Root 1: | 1.7071 | | | | |
| 10 | | Root 2: | 0.2929 | | | | |
| 11 | | | | | | | |

Follow this procedure to unlock a region of cells within a workbook:

1. Select the cells to be unlocked, cells C4:C6 in this example. (Remember: All cells are locked, by default.)
2. Unlock the cells by clicking the Lock Cell toggle button as illustrated in Figure 7.17: **Home tab → Cells group → Format drop-down menu → Lock Cell toggle button.** (Excel 2003: Format → Cells → Protection tab → Clear the Locked checkbox.)

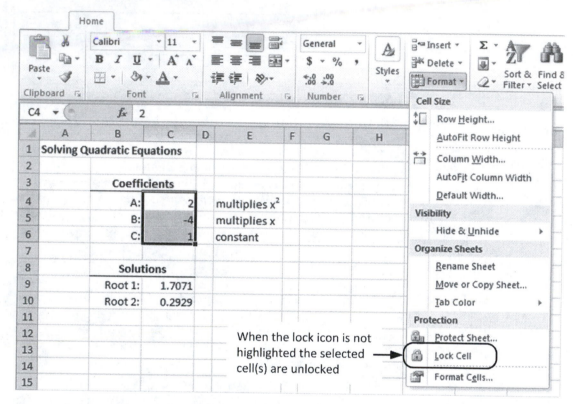

**Figure 7.17**
Unlocking selected cells.

Whether the cells are locked or unlocked makes no difference until the worksheet is protected. To activate worksheet protection, follow these steps:

1. Activate worksheet protection using one of these methods:
   a. Use Ribbon options: **Home tab → Cells group → Format drop-down menu → Protect Sheet . . . option.**
   b. Right-click on the worksheet tab and select **Protect Sheet . . .**
   c. Excel 2003: Tools → Protection → Protect Sheet.

The Protect Sheet dialog box will be displayed (Figure 7.18).

2. Verify that the **Protect worksheet and contents of locked cells** box is checked.
3. Enter a password if you wish. If you enter a password:
   • You will be prompted to confirm it.
   • You will need to use this password to make changes to the worksheet in the future.

**Figure 7.18**
The Protect Sheet
dialog box.

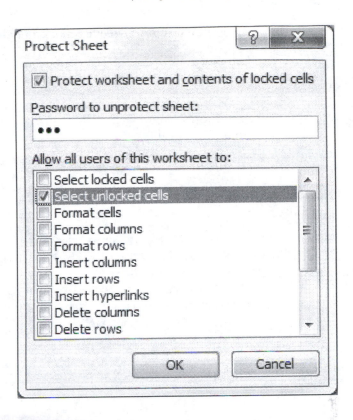

4. In the **Allow all users of this worksheet to** list:
   a. Check the box labeled **Select unlocked cells.**
   b. Clear all other check boxes.
5. Click **OK** to close the Protect Sheet dialog box. (You will be prompted to confirm the password if you entered one.)

After protecting the worksheet you can enter values for the coefficients in cells C4:C6 and the calculated results will change. However, you cannot select or modify any other cells.

You might want to allow users to **Select locked cells** as well (Figure 7.18). There are pros and cons to this:

- PRO: Users can click on the cells containing formula to verify the math.
- CON: Users can get frustrated when they can select cells, but are not allowed to modify them.

## 7.6 IMPORTING AND EXPORTING DATA FROM EXTERNAL FILE FORMATS

One problem of working as part of a team is that team members may use different application software products. In addition, a large and complex project may require the use of several software packages, such as Word, Access, and HTML documents. This implies that you may have to move data from one application to another in the process of completing a project.

Excel provides several methods for importing data. In this section, we will discuss two methods for importing external files: File Open and the Text Import Wizard.

- File Open—used when Excel has built-in methods for importing the type of file you are using.
- Text Import—used when you need to get basic text loaded into an Excel workbook.

Whenever Excel has built-in methods for importing the type of file you are using, you should use the File Open method to preserve as much of the original content as possible.

### 7.6.1 Importing by Using the File Open Option

A number of types of file formats may be imported directly into Excel. To import files, follow these steps:

1. Open the Open dialog box (see Figure 7.19):
   - Excel 2010: **File tab → Open button**
   - Excel 2007: **Office button → Open button**
   - Excel 2003: File → Open
2. Use the **All Excel Files** (Excel 2007: **Files of type**) drop-down menu (see Figure 7.19) to see a list of available file types (see Figure 7.20).
3. Select the file name extension of the file you want to import into Excel.

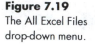

**Figure 7.19**
The All Excel Files drop-down menu.

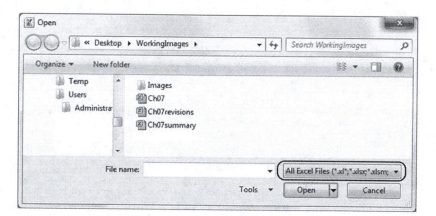

**Figure 7.20**
File types available in Excel 2010.

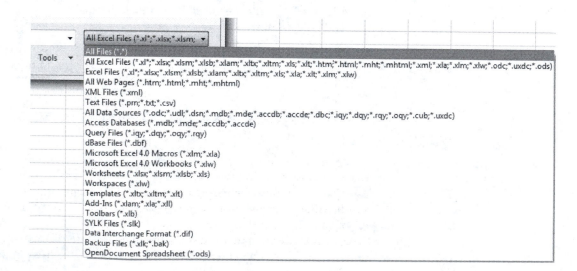

An external file type may be opened and viewed within Excel. If the file is modified, Excel will ask if you want to save the file in its original format or in Excel format.

- If the file is saved in Excel format, then the external program (e.g., dBase) will not be able to view the changes.
- If the file is saved in the external format (e.g., dBase), then some Excel formatting may be lost. For example, formulas and macros may not be translated into the external format.

Excel supports a wide range of file formats, but not all. When Excel does not support the file format used by your other software, sometimes a text file can be a useful intermediary.

### 7.6.2 Importing Text Data by Using the Text Import Wizard

Most applications will let you export data (using the **Save As** menu item) as tab- or *space-delimited text*. (A *delimiter* is a character that separates data values.) In addition, you may produce data from a computer program that you have written. In either case, the *Text Import Wizard* helps you align and import the data. When you import a text file you will get alphanumeric characters only. Formatting, colors, font size, etc., cannot be imported by using this method.

**Note:** A *Wizard* is a multi-step dialog box that leads the user through a multi-step process. Most of the Excel 2003 wizards disappeared with Excel 2007, but the Text Import Wizard is still used.

To demonstrate how to import a text file, we created a text file using a text editor. (Notepad in this example, but WordPad or Word could also be used.) The contents of the text file are shown in Figure 7.21. On each row the values have been separated by tabs.

**Figure 7.21**
Creating the MyTextFile.txt file using Notepad.

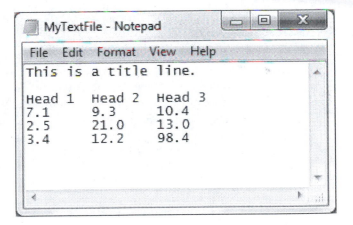

We begin the text import process by attempting to open the file in word.

1. Open the Open dialog box:
   a. Excel 2010: **File tab → Open button**
   b. Excel 2007: **Office button → Open button**
   c. Excel 2003: **File → Open**
2. Select **Text Files** in the **All Excel Files** (Excel 2007: **Files of type**) drop-down menu (see Figure 7.20).

3. Browse for the MyTextFile.txt file.
4. Click **Open** to attempt to open the file.

Excel will recognize that it is dealing with a text file, and automatically starts the Text Import Wizard. Step 1 is illustrated in Figure 7.22.

To import the file into Excel using the Text Import Wizard, follow these steps:

**Figure 7.22**
Text Import Wizard, Step 1 of 3.

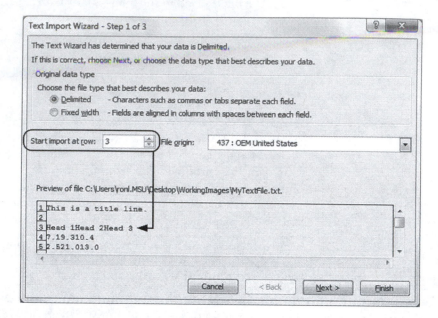

**Text Import Wizard, Step 1 of 3**

1. Indicate the type of data in the file. In this example, the data has been delimited with tabs.
   a. Delimited—values are separated by delimiters (such as the tabs we used).
   b. Fixed width—each column of values has the same number of characters.
2. Decide where you want Excel to begin importing the data. Many text files have some title text at the top of the file. Importing text files is cleaner when only the data is imported. We will skip the title line and begin the import at row 3.
3. Click **Next >** to go to Step 2 of 3, as shown in Figure 7.23.

**Text Import Wizard, Step 2 of 3**

1. Select the type(s) of delimiters used in the file. In this example, only tabs were used.
2. Click **Next >** to go to Step 3 of 3, as shown in Figure 7.24.

**Text Import Wizard, Step 3 of 3**

1. Verify the data format that will be used to import each column of data. In this example, the General format will be used. If you need to change a column data format:
   a. Select the column. (The selected column is shown in black with white text.)
   b. Choose a **Column data format** from the options list.
2. Click the **Finish** button to import the data into Excel.

**Figure 7.23**
Text Import Wizard,
Step 2 of 3.

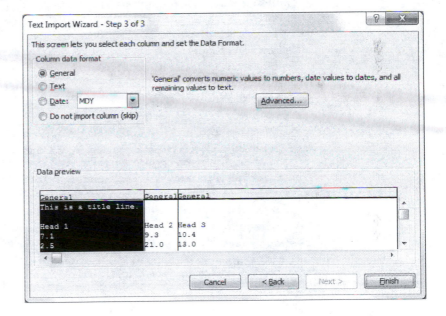

**Figure 7.24**
Text Import Wizard,
Step 3 of 3.

The imported data are shown in Figure 7.25. Notice that the title line was not imported because we asked that the import start in row 3 of the text file (in Step 1 of the Text Import Wizard).

Also, note that the file name shown in the Title bar in Figure 7.25 is still MyTextFile.txt. Excel does not automatically change the file extension to .xlsx. You should save the worksheet with the .xlsx extension to ensure that Excel features (e.g., formatting, formulas) are saved correctly. Excel will display a warning message if you attempt to save the worksheet with the .txt file extension.

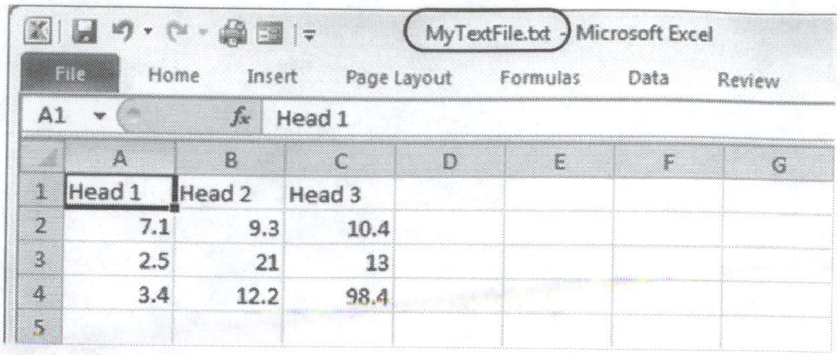

**Figure 7.25**
The imported text data.

## PROFESSIONAL SUCCESS—TEAM MEETING GUIDELINES

Read the following job descriptions, and notice what employers are looking for when hiring engineers.

**Mechanical Engineer**
We have an immediate need for an engineer to interface an engineering automation & optimization environment with a variety of CAD and CAE systems . . . B.S./M.S. in mechanical engineering . . . Good communication skills and *a strong interest in interacting with customers in problem-solving situations are a must.*

**Electrical Engineer**
Required degree: Perform audio subsystem validation to verify prototypes throughout the product-development program cycle. *Must work well in a team environment.*

**Aerospace Engineer**
Applicants selected may be subject to a government security investigation and meet eligibility requirements for access to classified information. *Must be a team player* and possess excellent written and oral communication skills.

**Industrial Engineer**
Investigate manufacturing processes in continuous-improvement environment, recommend refinements. Design process equipment to improve processes. Must be highly skilled at planning/managing and be *able to sell ideas to team members* and company management.

**Extrusion Engineer**
This manufacturer of fiber optics is seeking an Extrusion Engineer who can handle the majority of the technical issues in production. The successful candidate must be *able to work with other disciplines in a team atmosphere* of mutual support.

**Software Engineer**
Looking for a well-rounded software engineer with strong experience in object-oriented design and GUI development must be a highly motivated self-starter who *works well in a team environment.*

All of the positions listed were taken from actual job postings. What do the position announcements have in common? Teamwork! The ability of an engineer to work well in the team environment has as much to do with professional success as do scientific and technical skills. Few engineering accomplishments are produced in isolation.

Read the following expectations for the members of a professional team. You can start preparing for a career by practicing these skills.

### Decision Making

At first, attempt to make decisions by consensus. If that fails due to a single member who continually disagrees, then move to a consensus minus one approach. If consensus minus one fails, then move to a majority rule. Be aware that, as you lose the consensus of team members, the ability of the team to succeed is weakened. Time spent on gaining consensus by winning all team members over to an idea is time well spent!

### Confidentiality

Respect the other members' right to confidentiality. Lay ground rules about what material, if any, is to be treated confidentially. In the world of business and government contracts, you may be asked to sign a confidentiality agreement. These legal documents specify which of your employer's materials are protected from disclosure.

### Attention

Actively listen—ask questions or request clarification of other members' comments. Reflect a summary of important points that other team members have made. Give acknowledgement that you have understood. Try not to mentally rehearse what you are going to say while others are speaking.

### Preparation

Be adequately prepared for the meeting.

### Punctuality

Be on time. If you are 10 minutes late and there are six other members in the group, then you are wasting one person-hour of time!

### Ensure Active Contribution

If not all team members are contributing and actively participating, then something is wrong with the group process. Stop the meeting, and take time to get everyone involved before proceeding.

### Record Keeping

Someone on the team should be keeping records of team meetings.

### Flexibility

One of the aspects of working on a team is that you win some and you lose some. Not every one of your ideas will be accepted by the group. Be prepared to think of creative solutions that every team member can accept.

**Dynamics**

Help improve relationships among the team members. Do not dominate the meeting or let another member dominate the meeting. If this cannot be resolved within the group, enlist the help of an outside facilitator. A facilitator is a non-group member who does not take part in the content of the group issues. The facilitator exists to help smooth the group process. One of the roles of a facilitator is to prevent any single team member from dominating the others.

**Quorum**

Establish at the onset what a team quorum will be. Do not hold team meetings unless a quorum is present.

## USING TEAMWORK TO SOLVE A PROBLEM

Three students, Sara, Justin, and Allison, have been assigned a group project. Their assignment is to determine the energy required to pump water through a packed-bed filter for a local industrial facility (see Figure 7.26).

**Figure 7.26**
Water pumped through a packed-bed filter.

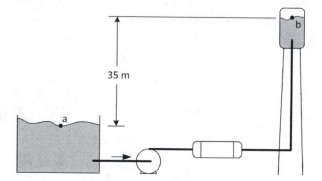

35 m

Here's what the team knows:

- Water flows at a rate of 400 liters per minute from a holding tank that is open to the atmosphere (pressure at point $a = 1$ atm $= 1.01 \times 10^5$ Pa) and into a centrifugal pump (25 HP, 60% efficiency).
- The water then flows through the packed-bed filter and into an elevated, pressurized storage tank (pressure at point $b = 3.00 \times 10^5$ Pa).
- The elevation change between the water level in the open tank (point $a$) and the water level in the pressurized tank (point $b$) is 35 m.
- The equation used to solve problems of this type is called the mechanical energy balance equation and is shown in Figure 7.27.
- The pump energy term on the left side of the equation represents the energy added to the system by the pump (expressed in *head* (height) of fluid—civil engineering style).
- The terms on the right side represent the energy required to change the pressure of the fluid, to lift the fluid from $H_a$ to $H_b$ (potential energy), to accelerate the fluid from $V_a$ to $V_b$, and to overcome friction ($F$). The group's task is to find $F$, which can be pretty significant with a packed bed in the flow line.

**Figure 7.27**
Mechanical energy balance.

$$\frac{\eta W_p}{g} = \frac{(P_b - P_a)}{\rho g} + (H_b - H_a) + \frac{1}{2}(\alpha_b V_b - \alpha_a V_a) + F$$

| pump energy term | pressure energy term | potential energy term | acceleration term | friction term |

The team devises the following strategy to solve the problem:

1. The first thing they do is to throw out the acceleration term, since the fluid velocity at the surface of each tank (at points *a* and *b*) will be very small. When those small velocities are squared, the acceleration term will be insignificant.
2. That leaves the pump energy term, the pressure energy term, and the potential energy term.
3. The students decide to each take one piece of the equation. They will each solve a piece of the larger problem in a separate workbook. Then they will combine their results by linking each workbook to find *F*. Sara takes the pump term, Justin takes the pressure term, and Allison takes the potential energy term.

**Sara's Part**

Sara's part is the most difficult. She is to find the pump energy term or $\dfrac{\eta W_p}{g}$.

$W_p$ is the energy per unit mass to the pump (usually from a motor), and $\eta W_p$ is the energy per unit mass from the pump to the fluid. Sara knows that the efficiency, $\eta$, is 60% or 0.60. But she doesn't know the energy per unit mass to the pump, $W_p$. She does know the power rating of the pump. Power is related to pump energy ($W_p$) through the mass flow rate through the pump:

$$Power = m_{flow} W_p$$

Mass flow rate is related to the stated volumetric flow rate through the fluid density,

$$m_{flow} = V_{flow}\rho.$$

The acceleration due to gravity $g = 9.8$ m/sec² at sea level. So Sara develops her worksheet, which looks like Figure 7.28. Sara saved her worksheet with the name PumpEnergy.xlsx.

The equations in Sara's worksheet are as follows:

| | | |
|---|---|---|
| C9: | =C2*745.7 | (the constant is 745.7 watts/HP) |
| C11: | =C4/1000/60 | (the 1,000 converts liters to m³ and the 60 converts minutes to seconds) |
| C16: | =C11*C12 | (density times volumetric flow rate) |
| C17: | =C9/C16 | (pump power rating divided by mass flow rate) |
| C19: | =C10*C17/C13 | $\left(\dfrac{\eta W_p}{g}\right)$ |

**Justin's Part**

Justin's spreadsheet is a bit simpler, since there are fewer conversions and calculations required. He is to find the pressure-energy term $= (P_b - P_a)/\rho \cdot g$. Justin's worksheet is shown in Figure 7.29. Justin saved his worksheet with the name PressureEnergy.xlsx.

**Figure 7.28**
Sara's worksheet finds the pump energy term.

| | A | B | C | D | E |
|---|---|---|---|---|---|
| 1 | Pump Term | | | | |
| 2 | | Power: | 25 | HP | |
| 3 | | η: | 0.60 | | |
| 4 | | V flow: | 400 | liters/min | |
| 5 | | ρ: | 1000 | kg/m$^3$ | |
| 6 | | g: | 9.8 | m/sec$^2$ | |
| 7 | | | | | |
| 8 | Converting to SI units | | | | |
| 9 | | Power: | 18642.5 | Joules/sec | |
| 10 | | η: | 0.60 | | |
| 11 | | V flow: | 0.00667 | m$^3$/sec | |
| 12 | | ρ: | 1000 | kg/m$^3$ | |
| 13 | | g: | 9.8 | m/sec$^2$ | |
| 14 | | | | | |
| 15 | Calculated Values | | | | |
| 16 | | m flow: | 6.667 | kg/sec | |
| 17 | | Wp: | 2796.4 | Joules/kg | |
| 18 | | | | | |
| 19 | | Pump Term: | 171.2 | J s$^2$ / m kg = m | |
| 20 | | | | | |

**Figure 7.29**
Justin's worksheet finds the pressure energy term.

| | A | B | C | D | E |
|---|---|---|---|---|---|
| 1 | Pressure Term | | | | |
| 2 | | P$_a$: | 1.01E+05 | Pa = n/m$^2$ | |
| 3 | | P$_b$: | 3.00E+05 | Pa = n/m$^2$ | |
| 4 | | ρ: | 1000 | kg/m$^3$ | |
| 5 | | g: | 9.8 | m/sec$^2$ | |
| 6 | | | | | |
| 7 | | Pressure Term: | 20.3 | J s$^2$ / m kg = m | |
| 8 | | | | | |

The equation in Justin's worksheet is

C7: =(C3-C2)/(C4*C5)

## Allison's Part

Allison's task is the easiest of all. She is to find the potential energy term $H_b - H_a$. Allison's worksheet is shown in Figure 7.30. Allison saved her worksheet with the name Potential.xlsx.

**Figure 7.30**
Allison's worksheet finds the potential energy term.

| | A | B | C | D | E |
|---|---|---|---|---|---|
| 1 | Potential Energy Term | | | | |
| 2 | | H$_a$: | 0 | m | |
| 3 | | H$_b$: | 35 | m | |
| 4 | | | | | |
| 5 | Potential Energy Term: | | 35 | m | |
| 6 | | | | | |

The equation in Allison's worksheet is

C5: C3-C2

## Combining the Results

Once each of the members had completed the assigned portion, the group got together and quickly finished the project. They created a summary workbook and created links to cells in their individual worksheets. Figure 7.31 shows the results.

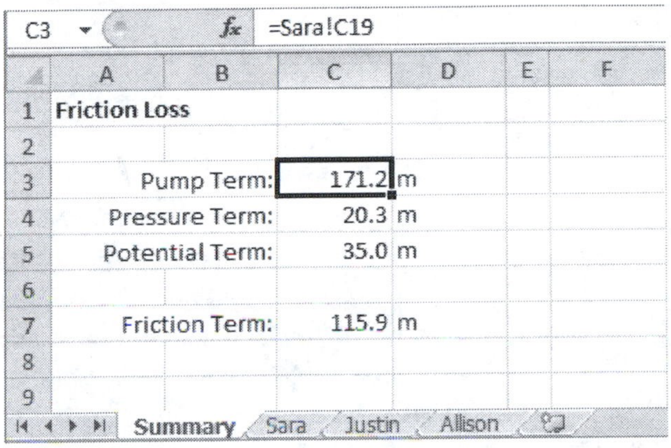

**Figure 7.31**
Summary worksheet used to find the friction term.

Sara combined all of their worksheets into a single workbook. The links that refer to other worksheets are shown next.

| | |
|------|-------------|
| C3: | =Sara!C19 |
| C4: | =Justin!C7 |
| C5: | =Allison!C5 |
| C7: | =C3-C4-C5 |

## NOTE

The external links can also refer to separate workbooks. The format for a remote cell reference is the workbook file name in square brackets, then worksheet name followed by an exclamation point, followed by the cell number.

[FileName]SheetName!CellRef

If a reference is made to a workbook in another directory, then the file name must contain the full pathname.

While this is a very simple example of using a spreadsheet for collaborating on group assignments, it does illustrate how easily the results from different members can be combined to complete a group assignment.

## KEY TERMS

change history
comment
delimiter
history tracking
locked cells
modify access
open access

open protection
password protection
protection
revision mark
shared workbook
sheet protection
space-delimited text

tab-delimited text
Text Import Wizard
tracking changes
write access
write protection

## SUMMARY

### Tracking Changes

*Identifying Yourself to Microsoft Office*

1. Open the Excel Options dialog box shown in Figure 7.4:
   - Excel 2010: **File tab → Options button**
   - Excel 2007: **Office button → Excel Options button**
   - Excel 2003: **Tools → Options**
2. Use the **General** panel, and the **Personalize your copy of Microsoft Office** section.
3. Enter your name in the User name box.
4. Click **OK** to close the Excel Options dialog box.

*Activating Change Tracking*

1. Open the Highlight Changes dialog box: **Review tab → Changes group → Track Changes drop-down menu → Highlight Changes . . . button**.
2. Check the **Track Changes while editing** box.
3. Check the **When** box and select **All**.
4. Check the **Who** box and select **Everyone**.
5. Check the **Where** box and select the entire worksheet.
6. Check the **Highlight changes on screen** box.
7. Click **OK**.

*Incorporating or Rejecting Revisions*

1. Open the Select Changes to Accept or Reject dialog box: **Review tab → Changes group → Track Changes drop-down menu → Accept/Reject Changes . . . button**.
2. Check the **When** box and select **Not Yet Reviewed**.
3. Check the **Who** box and select **Everyone**.
4. Check the **Where** box and select the entire worksheet.
5. Click **OK** to close the Select Changes to Accept or Reject dialog box. The Accept or Reject Changes dialog box will open.
6. Accept or reject each change as it is presented, or use the **Accept All** or **Reject All** buttons to process all changes at once.

**Adding a Comment to a Cell**

1. Select the cell that you want to comment on.
2. Click the **New Comment** button on the Review tab: **Review tab ➔ Comments group ➔ New Comment button**.
3. Type in the text of the comment, then click outside the comment box when finished.

**Sharing a Workbook**

1. Click the Share Workbook button to open the Share Workbook dialog box: **Review tab ➔ Changes group ➔ Share Workbook button**.
2. Select the Editing tab.
3. Check the box labeled **Allow changes by more than one user at the same time**.

**Setting Open Access Protection**

Open access protection requires a password to open a workbook.

1. Open the Save As dialog:
   - Excel 2010: **File tab ➔ Save As**
   - Excel 2007: **Office button ➔ Save As sub-menu ➔ Excel Workbook option**
   - Excel 2003: File ➔ Save As
2. Choose **Tools ➔ General Options . . .** from the Save As dialog box. The General Options dialog box will open.
3. Type a password in the box labeled **Password to open**.
4. Confirm the password.

**Setting Modify or Write Access Protection**

Modify or write access protection requires a password to save changes to a workbook.

1. Open the Save As dialog:
   a. Excel 2010: **File tab ➔ Save As**
   b. Excel 2007: **Office button ➔ Save As sub-menu ➔ Excel Workbook option**
   c. Excel 2003: File ➔ Save As
2. Choose **Tools ➔ General Options . . .** from the Save As dialog box. The General Options dialog box will open.
3. Type a password in the box labeled **Password to modify**.
4. Confirm the password.

**Worksheet Protection**

The general procedure for locking down a worksheet is as follows:

1. Unlock the cells that are to be available after the worksheet is locked.
2. Protect the worksheet.

**Unlocking Cells**

1. Select the cells to be unlocked.
2. Unlock the cells by clicking the Lock Cell toggle button: **Home tab ➔ Cells group ➔ Format drop-down menu ➔ Lock Cell toggle button**.

**Protect Worksheet**

1. Right-click on the worksheet tab and select **Protect Sheet . . .** The Protect Sheet dialog box will be displayed.
2. Verify that the **Protect worksheet and contents of locked cells** box is checked.
3. Enter a password if you wish.
4. Set desired access properties in the **Allow all users of this worksheet to** list.
5. Click **OK.**

**Text Import Wizard**

1. Open the Open dialog box (see Figure 7.19):
   a. Excel 2010: **File tab → Open button**
   b. Excel 2007: **Office button → Open button**
   c. Excel 2003: File → Open
2. Select **Text Files** in the **Files of type** drop-down menu.
3. Browse for the text file.
4. Click **Open** to attempt to open the file. The Text Import Wizard will open.

**Text Import Wizard, Step 1 of 3**

1. Indicate the type of data in the file.
   a. Delimited
   b. Fixed width
2. Decide on which row you want Excel to begin importing the data.
3. Click **Next >.**

**Text Import Wizard, Step 2 of 3**

1. Select the type(s) of delimiters used in the file.
2. Click **Next >.**

**Text Import Wizard, Step 3 of 3**

1. Verify the data format that will be used to import each column of data. If you need to change a column data format:
   a. Select the column.
   b. Choose a **Column data format** from the options list.
2. Click the **Finish** button to import the data into Excel.

## PROBLEMS

1. Practice merging workbooks with the following steps:
   a. Create a workbook with the data shown in Figure 7.3.
   b. Make two copies of the workbook.
   c. Make changes to each of the three documents.
   d. Merge the revised documents into a single document by opening the copy of the shared workbook into which you want to merge changes from another workbook file on disk.
   e. Merge the workbooks.
   f. Work through the process of accepting and rejecting the revisions.

2. Consider how passwords can be used to protect your Excel workbooks:
   a. Do the password-protection mechanisms discussed in this chapter prevent another student from making a copy of your Excel workbook? If so, which ones?
   b. Do any of the protection methods presented in the chapter prevent someone from printing your document without knowing the password? If so, which ones?

3. Turn sharing on and create a change history for a workbook. Then turn sharing off and see if the change history is actually deleted.

4. Set the change-history timer in the Share Workbook dialog box for one day. Wait more than 24 hours and see if the history really expires.

5. Create three workbooks, one for each of Sara's, Justin's, and Allison's parts of the Pump application in this chapter. Create a fourth, summary workbook that references cells in the other three workbooks and produces the final result (the friction term).

6. Create the quadratic equation solving tool shown in Figure 7.16. Then:
   • Unlock the cells containing the coefficients.
   • Protect the worksheet so that only to the cells containing the three coefficients are available.

**Figure 7.32**

Quadratic equation solving tool.

|  | A | B | C | D | E | F |
|---|---|---|---|---|---|---|
| 1 | Solving Quadratic Equations | | | | | |
| 2 | | | | | | |
| 3 | | Coefficients | | | | |
| 4 | | A: | 2 | | multiplies $x^2$ | |
| 5 | | B: | -4 | | multiplies x | |
| 6 | | C: | 1 | | constant | |
| 7 | | | | | | |
| 8 | | Solutions | | | | |
| 9 | | Root 1: | 1.7071 | | | |
| 10 | | Root 2: | 0.2929 | | | |
| 11 | | | | | | |

Use your worksheet to find solutions to the following quadratic equations:
   a. $2x^2 - 4x + 1 = 0$ (Solution shown in Figure 7.32.)
   b. $4x^2 - 5x + 1 = 0$
   c. $4x^2 - 8x - 5 = 0$

7. Delegating assignments to team members and then recombining the individual results to create the final product take some practice, but can work well if all do their part. As a practice problem, work with two other people to solve this problem: Find the total surface area of a hollow cylinder (Figure 7.33).

**Figure 7.33**

Schematic of the cylinder.

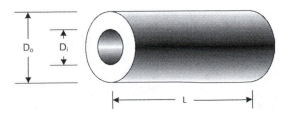

First, create the worksheet shown in Figure 7.34. There are no formulas in this worksheet, just labels (for now).

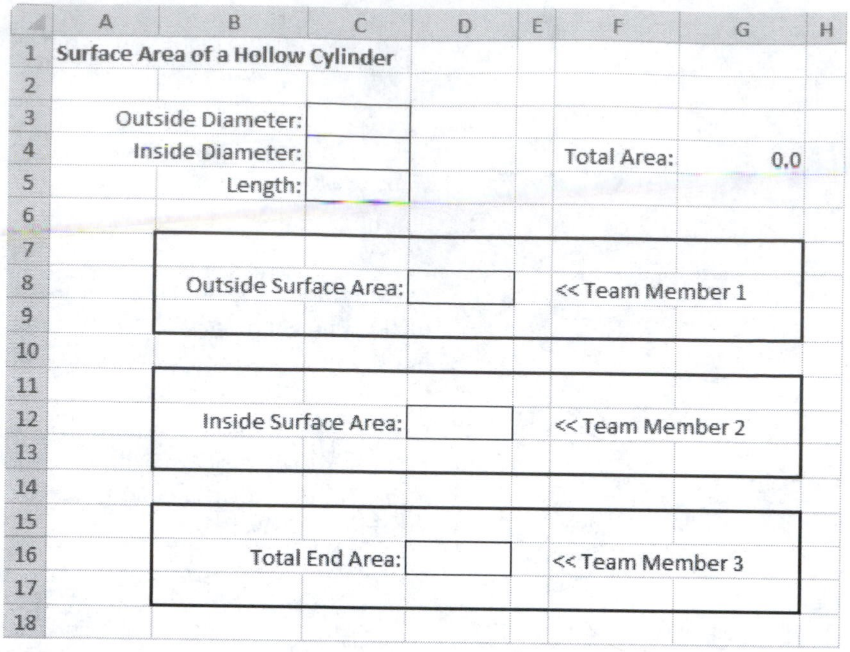

**Figure 7.34**
The labeled worksheet, before any formulas have been entered.

| | A | B | C | D | E | F | G | H |
|---|---|---|---|---|---|---|---|---|
| 1 | Surface Area of a Hollow Cylinder | | | | | | | |
| 2 | | | | | | | | |
| 3 | | Outside Diameter: | | | | | | |
| 4 | | Inside Diameter: | | | | Total Area: | 0.0 | |
| 5 | | Length: | | | | | | |
| 6 | | | | | | | | |
| 7 | | | | | | | | |
| 8 | | Outside Surface Area: | | | | << Team Member 1 | | |
| 9 | | | | | | | | |
| 10 | | | | | | | | |
| 11 | | | | | | | | |
| 12 | | Inside Surface Area: | | | | << Team Member 2 | | |
| 13 | | | | | | | | |
| 14 | | | | | | | | |
| 15 | | | | | | | | |
| 16 | | Total End Area: | | | | << Team Member 3 | | |
| 17 | | | | | | | | |
| 18 | | | | | | | | |

Once the labeled worksheet has been created, share the workbook so that it can be used by more than one person at a time. Next, save the workbook with three different names, one for each team member. Then, have each team member add his or her formula to their own worksheet, as follows:

Team Member 1:   Cell D8 = PI()*C3*C5

Team Member 2:   Cell D12 = PI()*C4*C5

Team Member 3:   Cell D16 = 2*PI()*((C3/2)^2 − (C4/2)^2)

**Note:** None of the formulas will return a numeric result yet, because the dimensions of the cylinder have not been specified. The team members can test their own portions with the following test dimensions, if desired.

Test dimensions:   $D_o = 3$   $D_i = 1$   $L = 7$

After each team member has added his or her formula to their own worksheet, merge the three worksheets. Then, in the merged workbook, enter the last formula to cell G3 (= D8 + D12 + D16) to add the areas calculated by each team member.

Finally, enter the test dimensions in cells C3:C5, and see if all of the pieces came together correctly. The result is shown in Figure 7.35.

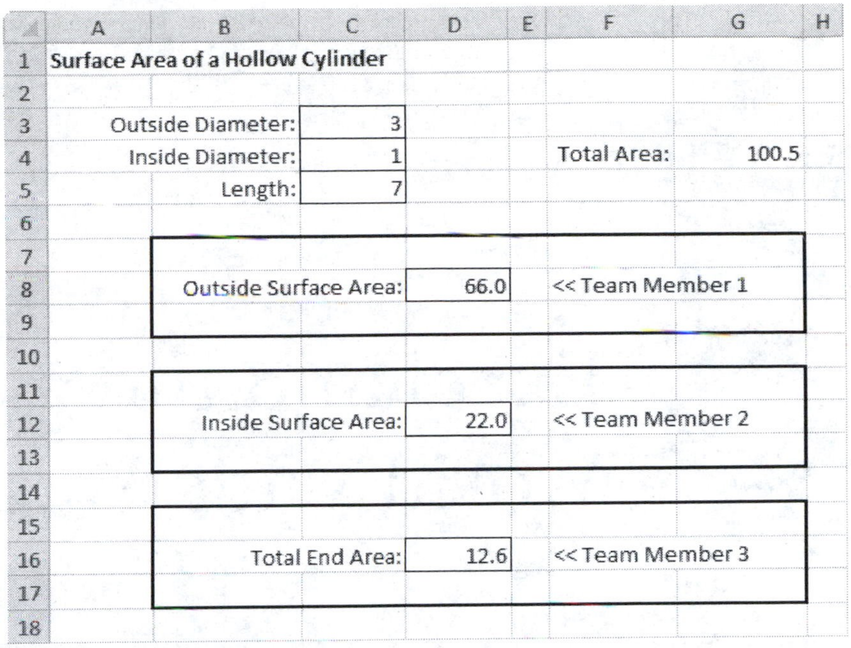

**Figure 7.35**
Combining workbooks to
calculate the surface area
of a hollow cylinder.

| | A | B | C | D | E | F | G | H |
|---|---|---|---|---|---|---|---|---|
| 1 | Surface Area of a Hollow Cylinder | | | | | | | |
| 2 | | | | | | | | |
| 3 | | Outside Diameter: | 3 | | | | | |
| 4 | | Inside Diameter: | 1 | | | Total Area: | 100.5 | |
| 5 | | Length: | 7 | | | | | |
| 6 | | | | | | | | |
| 7 | | | | | | | | |
| 8 | | Outside Surface Area: | | 66.0 | | << Team Member 1 | | |
| 9 | | | | | | | | |
| 10 | | | | | | | | |
| 11 | | | | | | | | |
| 12 | | Inside Surface Area: | | 22.0 | | << Team Member 2 | | |
| 13 | | | | | | | | |
| 14 | | | | | | | | |
| 15 | | | | | | | | |
| 16 | | Total End Area: | | 12.6 | | << Team Member 3 | | |
| 17 | | | | | | | | |
| 18 | | | | | | | | |

# 8

# Excel and the World Wide Web

## Section

## Objectives

*After reading this chapter, you should be able to perform the following tasks:*

- Access the World Wide Web from within an Excel worksheet.
- Retrieve files from HTTP servers into a local worksheet.

- Use the Web Query feature to import Excel data from the WWW.
- Create hyperlinks in a worksheet.
- Convert Excel documents to HTML.

## 8.1 ENGINEERING AND THE INTERNET

The *Internet* is one of the primary means of communication for scientists and engineers. Correspondence through electronic mail, the transfer of data and software *via* electronic file transfer, and research by using online search engines and databases are everyday tasks for engineers. The **World Wide Web** (*WWW.* or simply *Web*) is a collection of technologies for publishing, sending, and obtaining information by using the Internet. The Internet requires every engineering student to learn two new essential skills. First, every student must gain fluency in searching, locating, and retrieving relevant technical information from the Web. Second, every engineering student must learn how to post written documents to the Web. The ability to present technical results *via* the Web is an essential communication skill for modern engineers.

## PROFESSIONAL SUCCESS

The World Wide Web holds a wealth of information about your new profession. Take some time to visit the professional societies that represent your discipline. The following URLs represent a few of the national and international organizations that are online:

| | |
|---|---|
| Accreditation Board for Engineering and Technology (ABET) | www.abet.org |
| American Indian Science and Engineering Society (AISES) | www.aises.org |
| American Institute of Aeronautics and Astronautics (AIAA) | www.aiaa.org |
| American Institute of Chemical Engineers (AIChE) | www.aiche.org |
| American Society of Civil Engineers (ASCE) | www.asce.org |
| American Society for Engineering Education (ASEE) | www.asee.org |
| American Society of Mechanical Engineers (ASME) | www.asme.org |
| American Society of Naval Engineers (ASNE) | www.navalengineers.org |
| Engineers Without Borders (EWB) | www.ewb-usa.org |
| Institute of Electrical and Electronics Engineers (IEEE) | www.ieee.org |
| National Society of Black Engineers (NSBE) | www.nsbe.org |
| National Society of Professional Engineers (NSPE) | www.nspe.org |
| Society of Hispanic Professional Engineers (SHPE) | www.shpe.org |
| Society of Women Engineers (SWE) | www.swe.org |

## 8.2 ACCESSING THE WORLD WIDE WEB FROM WITHIN EXCEL

To access the Internet from within Excel, your computer must be connected to the Internet. If you are in a computer lab at school, then the computer may be connected to a local area network (LAN) through a network card. The LAN may or may not be connected to the Internet. Ask your lab manager or instructor for details. During the rest of this chapter, it is assumed that your computer is connected to the Internet.

Excel works well with the Web. Here are some basic Web-related tasks that will be covered in this chapter:

- Creating hyperlinks in Excel worksheets.
- Using Excel-related websites.
- Getting data from the Web into Excel.
  - Opening Excel files stored on the Web.
  - Copying and Pasting Web data into an Excel file.
  - Using a Query to obtain Web data.
- Saving an Excel worksheet as a Web document.

## 8.3 CREATING HYPERLINKS IN A WORKSHEET

A *hyperlink*, or simply *link*, can be thought of as a pointer to another location. When you click on a hyperlink, the contents at that location are immediately displayed. The hyperlink may point to:

- a cell in another worksheet in your Excel workbook
- another Excel worksheet on your local computer
- a document from another application, such as Microsoft Word
- a Web-based file stored on the other side of the planet
- a remote document that is retrieved from the World Wide Web using a transfer protocol

You are probably familiar with the "http" that precedes most Web addresses. It stands for *hypertext transfer protocol* and is one of the methods used to transfer information around the Web. The "http" in web addresses (aka *URLs*, or *uniform resource locators*) tells the browsers how to transfer the contents, but if you leave it out, modern browsers will determine an appropriate transfer protocol. Commonly used transfer protocols include:

- http—hypertext transfer protocol, for web pages
- ftp—file transfer protocol, for file exchange
- smtp—simple mail transfer protocol, for e-mail

### 8.3.1 Typing a Web Address in a Cell

The simplest way to insert a hyperlink in an Excel worksheet is to type a Web address in a cell. Excel will recognize the syntax and create the hyperlink. As an example, enter the Web address for Engineers Without Borders–USA, www.ewb-usa.org, in a cell, as shown in Figure 8.1. When you move the mouse over the hyperlink, the "finger" pointer is displayed. If you click on the link, a browser will open to display the Engineers Without Borders–USA website so that you can read about what engineering students are doing to improve people's lives around the world. (There are EWB organizations around the world; check out www.ewb-international.org to see a list of countries.)

**Figure 8.1**
Inserting a hyperlink in a cell.

### 8.3.2 Using the Insert Hyperlink Dialog Box

There is also an Insert Hyperlink dialog box to assist in creating a hyperlink. This can be helpful when you are not certain of the exact Web address, because you can search for the website you want using a browser. To use the Insert Hyperlink dialog box, follow these steps:

1. Right-click in the cell where the link will be placed. A pop-up menu will open.
2. Select **Insert Hyperlink . . .** from the pop-up menu. The Insert Hyperlink dialog box (Figure 8.2) will open.

**Figure 8.2**
The Insert Hyperlink
dialog box.

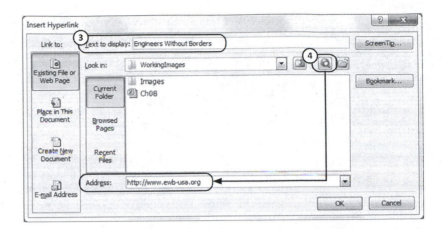

Alternate method: select the cell and use Ribbon options: **Insert tab → Link group → Insert Hyperlink button**.

3. In the **Text to display** field, enter the text that you want displayed as the link. If you leave this field blank, the actual web address will be displayed.

4. Click the **Browse the Web** button to search for the website you want to link to. The web address will appear in the **Address** field. Alternate method: simply type the web address in the **Address** field.

5. Click **OK** to close the Insert Hyperlink dialog box.

The link will be created in the cell, as illustrated in Figure 8.3.

**Figure 8.3**
The link created using
the Insert Hyperlink
dialog box.

| | A | B | C | D | E |
|---|---|---|---|---|---|
| 1 | | | | | |
| 2 | | | | | |
| 3 | | Engineers Without Borders | | | |
| 4 | | | | | |
| 5 | | | | | |

### 8.3.3 Links Within Your Excel Workbook

Another use for hyperlinks is providing quick access to other locations in the same workbook. For example, the worksheet shown in Figure 8.4 provides an overview of the anticipated profit on a project, but readers are almost certainly going to want to see how the total revenue and total cost values were calculated. Links provide a quick way for the reader to get to the detailed calculations.

**Figure 8.4**
Links to other locations in
the same workbook.

| | A | B | C | D | E | F |
|---|---|---|---|---|---|---|
| 1 | Project Summary Page | | | | | |
| 2 | | | | | Details... | |
| 3 | Total Revenues: | | $ 1,230,000 | | Revenue Data | |
| 4 | Total Costs: | | $  847,328 | | Cost Data | |
| 5 | | | | | | |
| 6 | Net Profit: | | $  382,672 | | | |
| 7 | | | | | | |

The workbook contains three worksheets, named Summary, Costs, and Revenues. To insert the link to the revenue data (on the Revenues worksheet), use these steps:

1. Right-click in cell E3, where the revenue data link will be placed. A pop-up menu will open.
2. Select **Insert Hyperlink . . .** from the pop-up menu. The Insert Hyperlink dialog box (Figure 8.5) will open.
3. Click the **Place in This Document button** in the **Link to:** list (indicated in Figure 8.5).
4. Enter *Revenue Data* in the **Text to Display** field.
5. Select the **Revenues** sheet in the **Or select a place in this document** list. The default is to link to cell A1 on the selected sheet, and we will leave that default unchanged.
6. Click **OK** to close the Insert Hyperlink dialog box.

The result is shown in Figure 8.4. When you click on the Revenue Data link, Excel displays the Revenues sheet as shown in Figure 8.6.

**Figure 8.5**

The Insert Hyperlink dialog box when inserting a link within a worksheet.

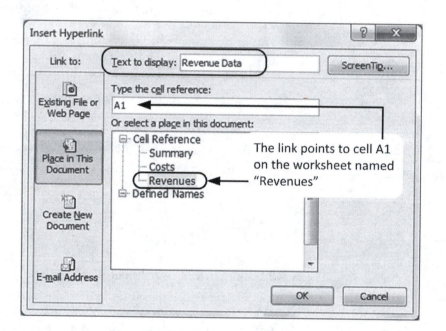

**Figure 8.6**

The result of following the Revenue Data link.

| | A | B | C | D | E | F | G |
|---|---|---|---|---|---|---|---|
| 1 | Revenues | | | | | | |
| 2 | | Item # | Description | Number | Price | Revenue | |
| 3 | | 1 | Small basic widgets | 6,438 | $ 121.50 | $ 782,217 | |
| 4 | | 2 | Small deluxe widgets | 418 | $ 165.50 | $ 69,179 | |
| 5 | | 3 | Large basic widgets | 1,298 | $ 265.38 | $ 344,463 | |
| 6 | | 4 | Large deluxe widgets | 110 | $ 310.37 | $ 34,141 | |
| 7 | | | | | | | |
| 8 | | | | | Total Revenue: | $1,230,000 | |
| 9 | | | | | | | |

## 8.4 USING WEBSITES RELATED TO EXCEL

Excel is such a commonly used program that there are a huge number of websites related to its use. These fall into two basic categories:

1. How to use the Excel program.
2. How to use Excel to accomplish some task.

Microsoft maintains a section of web pages specifically for Excel users. The primary page for Microsoft is located at the following URL:

http://www.microsoft.com

Within the Microsoft site are Excel tutorials, product information, and a number of free add-ins, patches, and templates. It is important to periodically check for updates to your programs, since security problems are frequently found in most popular applications.

There are many other websites available that provide instruction on using Excel for a myriad of tasks. A good way to search for these sites is to enter a topic in a search engine such as Google.[1] For example, you could search for "hyperlinks in Excel" and see results something like those shown in Figure 8.7.

**Figure 8.7**

Partial results of a Google search on "hyperlinks in Excel."

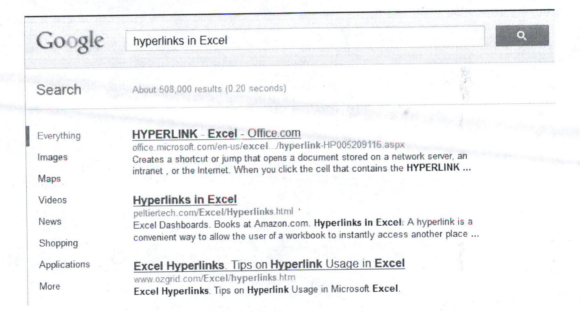

## 8.5 USING WEB DATA IN EXCEL

The Web is a huge source of information for engineers. The way you access that information depends on how the information is stored.

- Data in tables on web pages can be copied and pasted into Excel.
- Excel workbooks can be accessed over the Web.
- Web queries can be used to retrieve data over the Web.

We'll consider each of these ways of accessing data from the Web.

---

[1] Google is a trademark of Google Inc.

## 8.5.1 Copying and Pasting Web Data into Excel

A lot of engineering data is published to the Web for others to use. For example, the U.S. Department of Energy's Energy Information Administration (EIA) provides a wealth of data on energy-related topics. For example, oil price information is available at:

http://www.eia.gov/totalenergy/data/monthly/

A portion of the website is shown in Figure 8.8.

**Figure 8.8**
EIA data on crude oil prices (partial).

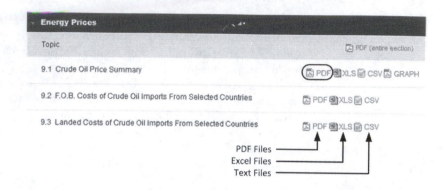

Notice that the data is provided in three different formats:

- PDF files
- Excel files
- Text files

We'll start by opening the PDF page containing a crude oil price summary (indicated in Figure 8.8). This data set lists crude oil prices since 1968. The table is quite large, and only a portion of it is shown in Figure 8.9.

**Figure 8.9**
Crude oil cost data (part of Web table).

www.eia.gov/totalenergy/data/monthly/pdf/sec9_3.pdf

### Table 9.1 Crude Oil Price Summary
(Dollars[a] per Barrel)

| | Domestic First Purchase Price[c] | F.O.B. Cost of Imports[d] | Landed Cost of Imports[e] | Refiner Acquisition Cost[b] | | |
|---|---|---|---|---|---|---|
| | | | | Domestic | Imported | Composite |
| 1973 Average | 3.89 | [f] 5.21 | [f] 6.41 | [E] 4.17 | [E] 4.08 | [E] 4.15 |
| 1975 Average | 7.67 | 11.18 | 12.70 | 8.39 | 13.93 | 10.38 |
| 1980 Average | 21.59 | 32.37 | 33.67 | 24.23 | 33.89 | 28.07 |
| 1985 Average | 24.09 | 25.84 | 26.67 | 26.66 | 26.99 | 26.75 |
| 1990 Average | 20.03 | 20.37 | 21.13 | 22.59 | 21.76 | 22.22 |
| 1995 Average | 14.62 | 15.69 | 16.78 | 17.33 | 17.14 | 17.23 |
| 1996 Average | 18.46 | 19.32 | 20.31 | 20.77 | 20.64 | 20.71 |
| 1997 Average | 17.23 | 16.94 | 18.11 | 19.61 | 18.53 | 19.04 |
| 1998 Average | 10.87 | 10.76 | 11.84 | 13.18 | 12.04 | 12.52 |
| 1999 Average | 15.56 | 16.47 | 17.23 | 17.90 | 17.26 | 17.51 |
| 2000 Average | 26.72 | 26.27 | 27.53 | 29.11 | 27.70 | 28.26 |
| 2001 Average | 21.84 | 20.46 | 21.82 | 24.33 | 22.00 | 22.95 |
| 2002 Average | 22.51 | 22.63 | 23.91 | 24.65 | 23.71 | 24.10 |
| 2003 Average | 27.56 | 25.66 | 27.69 | 29.82 | 27.71 | 28.53 |
| 2004 Average | 36.77 | 33.75 | 36.07 | 38.97 | 35.90 | 36.98 |
| 2005 Average | 50.28 | 47.60 | 49.29 | 52.94 | 48.86 | 50.24 |
| 2006 Average | 59.69 | 57.03 | 59.11 | 62.62 | 59.02 | 60.24 |
| 2007 Average | 66.52 | 66.36 | 67.97 | 69.65 | 67.04 | 67.94 |
| 2008 Average | 94.04 | 90.32 | 93.33 | 98.47 | 92.77 | 94.74 |

**Figure 8.10**
After pasting the table into Excel and graphing.

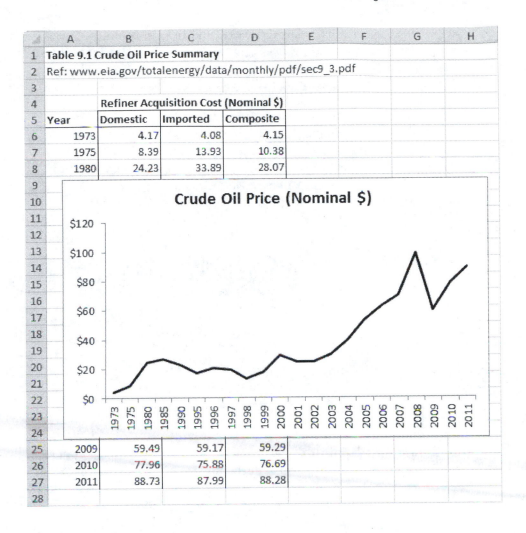

|   | A | B | C | D | E | F | G | H |
|---|---|---|---|---|---|---|---|---|
| 1 | Table 9.1 Crude Oil Price Summary | | | | | | | |
| 2 | Ref: www.eia.gov/totalenergy/data/monthly/pdf/sec9_3.pdf | | | | | | | |
| 3 | | | | | | | | |
| 4 | | Refiner Acquisition Cost (Nominal $) | | | | | | |
| 5 | Year | Domestic | Imported | Composite | | | | |
| 6 | 1973 | 4.17 | 4.08 | 4.15 | | | | |
| 7 | 1975 | 8.39 | 13.93 | 10.38 | | | | |
| 8 | 1980 | 24.23 | 33.89 | 28.07 | | | | |

**Crude Oil Price (Nominal $)**

| 25 | 2009 | 59.49 | 59.17 | 59.29 | | | | |
| 26 | 2010 | 77.96 | 75.88 | 76.69 | | | | |
| 27 | 2011 | 88.73 | 87.99 | 88.28 | | | | |

To copy the data to the Windows clipboard:

1. Click inside the table.
2. Select the entire table using browser options **Edit → Select All** (or the shortcut, **Ctrl + A**).
3. Use browser menu options **Edit → Copy** (or the shortcut, **Ctrl + C**).

To paste the data into Excel:

1. Click in the cell that will hold the top-left corner of the table.
2. Paste the table into Excel using Ribbon Options **Home tab → Clipboard group → Paste button** (or the shortcut, **Ctrl + V**).

**Note:** This approach may paste all information in a row into a single cell. When that happens, use Excel's **Paste Options** and **Use Text Import Wizard . . .**

The result of the paste operation is shown in Figure 8.10. To demonstrate that the pasted values can be used in Excel, the nominal (not adjusted for inflation) price of crude oil has been plotted using an Excel Line chart.

Some data tables paste into Excel more cleanly than others. Here are some pointers to improve your ability to copy and paste values into Excel:

- Select the entire table before copying.
- Use **Paste Special**, **As Text**, and **Use Text Import Wizard** . . . if needed.

However, the Energy Information Administration provides a better option: downloading the data as Excel files.

### 8.5.2 Downloading Excel Files from the Web

If the data you need is available as an Excel file, then you can download the file and open it in Excel. As an example, let's open the Excel file (actually, it's an .asp file, but Excel can open it) labeled Fuel Ethanol and Biodiesel Overview, 1981–2007 in the EIA website (indicated in Figure 8.11, and located at www.eia.doe.gov/emeu/aer/renew.html).

**Figure 8.11**

EIA data on renewable energy (partial).

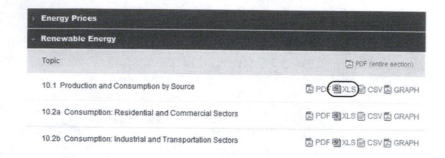

When you click on the link, your browser will likely display a dialog box asking what to do with the file: save it, or open it in Excel. Open the file in Excel. The result is shown in Figure 8.12. A chart of renewable energy use since 1973, created from the EIA data, is shown in Figure 8.13.

**Figure 8.12**

Partial display of downloaded Excel file.

| | A | C | D | E | F |
|---|---|---|---|---|---|
| 1 | U.S. Energy Information Administration | | | | |
| 2 | *February 2012 Monthly Energy Review* | | | | |
| 3 | | | | | |
| 9 | Table 10.1 Renewable Energy Production and Consumption by Source | | | | |
| 11 | | | | | |
| 12 | **Year** | **Total Biomass Energy** | **Total Renewable** | **Hydroelectric Power** | **Geothermal Energy** |
| 13 | | **(Trillion Btu)** | **(Trillion Btu)** | **(Trillion Btu)** | **(Trillion Btu)** |
| 26 | 1973 Total | 1529.1 | 4410.9 | 2861.4 | 20.4 |
| 39 | 1974 Total | 1539.7 | 4741.9 | 3176.6 | 25.6 |
| 52 | 1975 Total | 1498.7 | 4687.1 | 3154.6 | 33.8 |
| 65 | 1976 Total | 1713.4 | 4727.2 | 2976.3 | 37.5 |
| 78 | 1977 Total | 1838.3 | 4209.0 | 2333.3 | 37.4 |
| 91 | 1978 Total | 2037.6 | 5005.4 | 2937.0 | 30.9 |
| 104 | 1979 Total | 2151.9 | 5122.9 | 2930.7 | 40.3 |
| 117 | 1980 Total | 2475.5 | 5428.3 | 2900.1 | 52.7 |
| 130 | 1981 Total | 2596.3 | 5413.7 | 2758.0 | 59.4 |
| 143 | 1982 Total | 2663.5 | 5979.6 | 3265.6 | 50.6 |

**Figure 8.13**
A plot of renewable energy use since 1973, created using data from the EIA website.

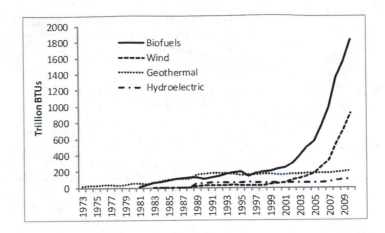

### 8.5.3 Using a Web Query to Retrieve Web Data

A *Web Query* retrieves data from an external source over the web and places the data in a local Excel worksheet. In this example, we will use a Web Query to get current currency exchange rates from the file rates.asp at http://moneycentral.msn.com. This is one of several sample queries that are provided with a standard Excel installation.

To run a Web Query, follow these steps:

1. Open a new blank Excel workbook.
2. Open the Existing Connections dialog box (Figure 8.14) using Ribbon options: **Data tab → Get External Data group → Existing Connections button**. (Excel 2003: Open the Select Data Source dialog box using Data → Import External Data → Import Data.)
3. Select **MSN MoneyCentral Investor Currency Rates** (as shown in Figure 8.14).

**Figure 8.14**
The Existing Connections dialog box.

4. Click **Open**. The Import Data dialog box will open, as shown in Figure 8.15.
5. Choose a location for the imported query results (cell A1 in the existing worksheet by default).
6. Click **OK** to retrieve the data.

The results are shown in Figure 8.16.

**Figure 8.15**
The Import Data dialog box.

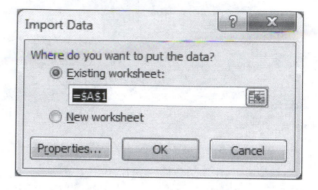

**Figure 8.16**
The query results.

| | A | B | C | D |
|---|---|---|---|---|
| 1 | **Currency Rates Provided by MSN Money** | | | |
| 2 | Click here to visit MSN Money | | | |
| 3 | | | | |
| 4 | **Name** | **In US$** | **Per US$** | |
| 5 | Argentine Peso to US Dollar | 0.22962 | 4.355 | |
| 6 | Australian Dollar to US Dollar | 1.0591 | 0.944 | |
| 7 | Bahraini Dinar to US Dollar | 2.6488 | 0.378 | |
| 8 | Bolivian Boliviano to US Dollar | 0.14451 | 6.92 | |
| 9 | Brazilian Real to US Dollar | 0.55393 | 1.805 | |
| 10 | British Pound to US Dollar | 1.5838 | 0.631 | |
| 11 | Canadian Dollar to US Dollar | 1.0078 | 0.992 | |

A Web Query is a formatted text file. The contents of the MSN query used to access the data in the previous example are displayed in Figure 8.17. The effect of executing the query is to access the web server at the URL and execute the named command. The results are returned and displayed in your local Excel worksheet.

You can ask Excel to refresh the data periodically. This could be important if your business handled a lot of different currencies. To set the query refresh period:

1. Click the **Properties** button on the Import Data dialog box. This opens the External Data Range Properties dialog box shown in Figure 8.18.
2. Check the **Refresh every** box and set the interval to 60 minutes, or whatever refresh interval is needed.
3. Click **OK** to close the dialog box.

As long as you have your worksheet open and you are connected to the Internet, the Web Query will automatically run every 60 minutes and update the currency rates.

**Figure 8.17**
The query used to retrieve the currency values.

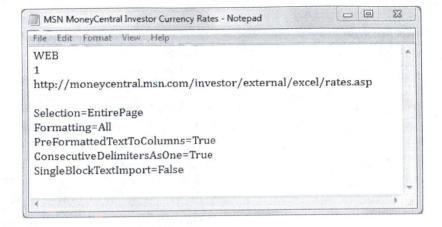

```
WEB
1
http://moneycentral.msn.com/investor/external/excel/rates.asp

Selection=EntirePage
Formatting=All
PreFormattedTextToColumns=True
ConsecutiveDelimitersAsOne=True
SingleBlockTextImport=False
```

**Figure 8.18**
The External Data Range Properties dialog box.

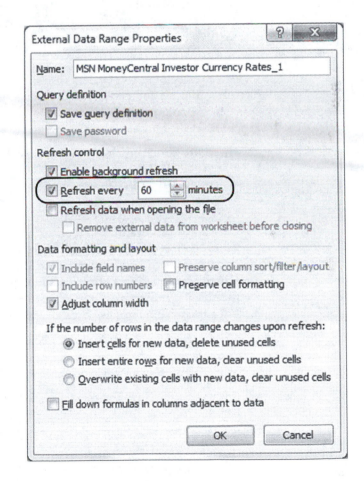

## 8.6 SAVING AN EXCEL WORKBOOK AS A WEB PAGE

Once you have created an Excel workbook, you can save it as an HTML *web page*. Not all Excel features are supported by *HTML (hypertext markup language)*, so you want to save your work as an Excel file (.xlsx or .xlsm) first, and then create the web page.

The basic steps involved are:

1. Choose a web page format.
   - Web Page (*.htm, *.html)
   - Single File Web Page (*.mht; *.mhtml)

   The **Web Page format** creates a web page of your workbook, plus a folder containing supporting materials (graphics, mostly). This format minimizes the size of the web page, but requires that the supporting folder be maintained.

   The **Single File Web Page format** (not available in Excel 2003) puts everything needed to display the web page in a single file. It is convenient because you don't have to worry about the supporting folder, but the file can be large.

2. Choose a file name.
3. Assign a page title.
4. Select whether to create a web page of the entire workbook, or just the current worksheet.
5. Create the web page.

These basic steps will be demonstrated using the workbook shown in Figure 8.4 that presents cost and revenue data for a project. We will create a web page for the entire workbook.

To save this workbook as a web page:

1. Open the Save As dialog box (Figure 8.19):
   - Excel 2010: **File tab → Save As button**
   - Excel 2010: **Office button → Save As sub-menu → Other Formats button**
   - Excel 2003: File → Save as Web Page
2. Choose the **Save as type** format. Standard **Web Page (*.htm, *.html)** format was selected in this example.
3. Enter the file name **ProjData** in the **File name** field.
4. Use the **Change Title . . .** button to assign the title **Project Data** to the web page.
5. Use the **Save** option **Entire Workbook** to create a web page of the entire Excel workbook.
6. Click the **Save** button to save the workbook as a web page.

Alternative: The **Publish** button allows you to republish the web page each time the Excel file is saved.

The results of saving the workbook as a web page include:

- A web page with the file name **ProjData.htm**, shown in Figure 8.20.
- A folder named **ProjData_Files** containing an .htm file for each of the three worksheets in the workbook, plus the tab strip (shown in Figure 8.20) and supporting materials.

**Figure 8.19**
The Save As dialog box,
filled in.

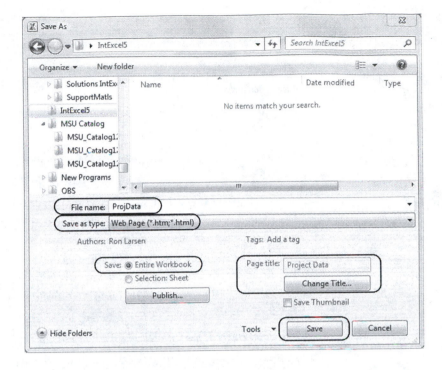

**Figure 8.20**
The web page created
from the Excel workbook.

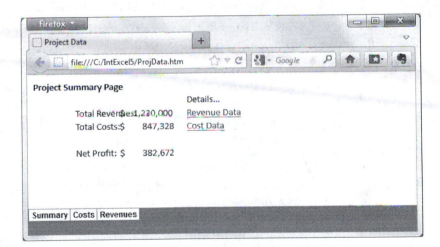

## SUMMARY

### Creating Hyperlinks

*Two Methods*

1. Type a web address into a cell.
2. Use the Insert Hyperlink dialog box.

*Using the Insert Hyperlink Dialog Box to Create an External Link*

1. Right-click in the cell where the link will be placed.
2. Select **Insert Hyperlink . . .** from the pop-up menu.
3. In the **Text to display** field, enter the text that you want displayed as the link. If you leave this field blank, the actual web address will be displayed.
4. Click the **Browse the Web** button to search for the website you want to link to. The web address will appear in the **Address** field.

   Alternate method: Simply type the web address in the **Address** field.
5. Click **OK** to close the Insert Hyperlink dialog box.

*Using the Insert Hyperlink Dialog Box to Create a Link Within Your Excel Workbook*

1. Right-click in the cell where the link will be placed.
2. Select **Insert Hyperlink . . .** from the pop-up menu.
3. Click the **Place in This Document** button in the **Link to:** list.
4. Enter the link label text in the **Text to Display** field.
5. Select the link location (sheet and/or cell location).
6. Click **OK** to close the Insert Hyperlink dialog box.

### Using Data from the Web

*Common Ways to Access Web Data:*

- Data in tables on web pages can be copied and pasted into Excel.
- Excel workbooks can be accessed over the Web.

*Copying and Pasting PDF Web Data into Excel*

*To copy the data to the Windows clipboard:*

1. Click inside the PDF table.
2. Select the entire table using browser options **Edit → Select All** (or the shortcut, **Ctrl + A**).
3. Use browser menu options **Edit → Copy** (or the shortcut, **Ctrl + C**).

*To paste the data into Excel:*

1. Click in the cell that will hold the top-left corner of the table.
2. Paste the table into Excel using Ribbon Options **Home tab → Clipboard group → Paste button** (or the shortcut **Ctrl + V**).

   **Note:** This approach may paste all information in a row into a single cell. When that happens, use Excel's **Paste Options** and **Use Text Import Wizard . . .**

*Downloading Excel Files from the Web*

1. Click the Web link to the Excel file.
2. Indicate whether the file should be saved or opened (if asked).

**Saving a Worksheet as a Web Page**

1. Open the Save As dialog box:
   a. Excel 2010: **File tab → Save As button**
   b. Excel 2010: **Office button → Save As sub-menu → Other Formats button**
   c. Excel 2003: File → Save as Web Page
2. Choose the **Save as type** format (Web Page, or Single File Web Page).
3. Enter the file name in the **File name** field.
4. Use the **Change Title . . .** button to assign a title to the web page.
5. Use the **Save** options to choose to save the entire workbook or just the current worksheet.
6. Click the **Save** button to save the workbook as a web page.

**Web Format Options**

- The **Web Page format** creates a web page of your workbook, plus a folder containing supporting materials (graphics, mostly). This format minimizes the size of the web page, but requires that the supporting folder be maintained.
- The **Single File Web Page format** (not available in Excel 2003) puts everything needed to display the web page in a single file. It is convenient because you don't have to worry about the supporting folder, but the file can be large.

## PROBLEMS

1. Create a workbook named SI Units.xlsx. Enter the label SI Base Unit Definitions in a cell, and create a hyperlink from this cell to:

   http://physics.nist.gov/cuu/Units/current.html

2. Use the Help feature to read about the HYPERLINK function (another method of creating a hyperlink, not covered in this chapter). One advantage of using the HYPERLINK function is that the link can depend on a conditional expression. Create an *IF* expression that links to www. mozilla .com if cell *A1 = Firefox* and links to www.microsoft.com if cell *A1 = Explorer*.

3. One use of hyperlinks in Excel worksheets is to provide quick access to online reference materials. As an example, perform a Google® search to find the equation used to calculate the surface area of a cone. Then finish the worksheet shown in Figure 8.21 by using the equation to calculate the surface area and adding a hyperlink to the site that provided the equation.

4. The British have long used the *stone* as a unit of mass, and it continues to be used in everyday speech, even though the United Kingdom switched to the SI system of units many years ago. Search the Web to find out how many

**Figure 8.21**
Partially completed
worksheet for computing
the surface area of a cone.

| | A | B | C | D | E |
|---|---|---|---|---|---|
| 1 | Surface Area of a Cone | | | | |
| 2 | | | | | |
| 3 | | Base Radius: | 2 | cm | |
| 4 | | Height: | 6 | cm | |
| 5 | | | | | |
| 6 | | Surface Area: | | cm$^2$ | |
| 7 | | | | | |
| 8 | | Link to Reference: | | | |
| 9 | | | | | |

pounds are equal to 1 stone, then create a worksheet that accomplishes the following:

**a.** Computes the mass of a 150-pound individual, in stones.

**b.** Provides a hyperlink to the website that was used to find out how pounds and stones are related.

5. An online search has revealed that a plasma television can be purchased from Store A for 650 US$ + 50 US$ s/h, or from Store B for 700 C$ + 63 C$ s/h. After a bit of searching, you figure out that US$ stands for U.S. dollars, C$ means Canadian dollars, and s/h stands for shipping and handling. Create a worksheet that performs these tasks:

**a.** Computes the total cost of each system, and converts all costs to the same currency, either U.S. or Canadian.

**b.** Provides a hyperlink to a website that provides currency exchange rates, such as http://www.xe.com/ucc/.

Which store has the better price?

6. After searching an online auction site for a new rug, you have found exactly the item you want. You submit a bid and win the item, and then notice that the rug is in Pakistan and you just agreed to pay 22,000 Pakistani rupees. Furthermore, after the auction you learn that the seller is happy to ship to your country, at an additional cost of 34,000 rupees.

Create a worksheet to compute exactly what the rug is going to cost you, including shipping. Provide a hyperlink to the website you used to find the exchange rate between Pakistani rupees and your currency.

# Commonly Used Functions

| | |
|---|---|
| ABS($n$) | Returns the absolute value of a number |
| AND($a, b, \ldots$) | Returns the logical AND of the arguments (TRUE if all arguments are TRUE, otherwise FALSE) |
| ASIN($n$) | Returns the arcsine of $n$ in radians |
| AVEDEV($n1, n2, \ldots$) | Returns the average of the absolute deviations of the arguments from their mean |
| AVERAGE($n1, n2, \ldots$) | Returns the arithmetic mean of its arguments |
| BIN2DEC($n$) | Converts a binary number to decimal |
| BIN2HEX($n$) | Converts a binary number to hexadecimal |
| BIN2OCT($n$) | Converts a binary number to octal |
| CALL($\ldots$) | Calls a procedure in a DLL or code resource |
| CEILING($n, sig$) | Rounds a number $n$ up to the nearest integer (or nearest multiple of significance, $sig$) |
| CHAR($n$) | Returns the character represented by the number $n$ in the computer's character set |
| CHIDIST($x, df$) | Returns the one-tailed probability of the chi-squared distribution, using degrees of freedom ($df$) |

| | |
|---|---|
| CLEAN(*text*) | Removes all nonprintable characters from *text* |
| COLUMN(*ref*) | Returns the column number of a reference |
| COLUMNS(*ref*) | Returns the number of columns in a reference |
| COMBIN(*n, r*) | Returns the number of combinations of *n* items, choosing *r* items |
| COMPLEX(*real, imag, suffix*) | Converts real and imaginary coefficients into a complex number |
| CONCATENATE (*str1, str2, . . .*) | Concatenates the string arguments |
| CORREL(**A1, A2**) | Returns the correlation coefficients between two data sets |
| COS(*n*) | Returns the cosine of an angle |
| COUNTBLANK(*range*) | Counts the number of empty cells in a specified range |
| DEC2BIN(*n, p*) | Converts the decimal number *n* to binary, using *p* places (or characters) |
| DELTA(*n1, n2*) | Tests whether two numbers are equal |
| ISERROR(*v*) | Returns TRUE if value *v* is an error |
| ISNUMBER(*v*) | Returns TRUE if value *v* is a number |
| FACT(*n*) | Returns the factorial of *n* |
| FORECAST(*x, known x's, known y's*) | Predicts a future value based on a linear trend |
| LN(*n*) | Returns the natural logarithm of *n* |
| MDETERM(**A**) | Returns the matrix determinant of array **A** |
| MEDIAN(*n1, n2, . . .*) | Returns the median of its arguments |
| MOD(*n, d*) | Returns the remainder after *n* is divided by *d* |
| OR(*a, b, . . .*) | Returns the logical OR of its arguments (TRUE if any argument is TRUE, FALSE if all arguments are FALSE) |
| PI() | Returns the value of *pi* to 15 digits of accuracy |
| POWER(*n, p*) | Returns the value of *n* raised to the power of *p* |
| PRODUCT(*n1, n2, . . .*) | Returns the product of its arguments |
| QUOTIENT(*n, d*) | Returns the integer portion of *n* divided by *d* |
| RADIANS(*d*) | Converts degrees to radians |
| RAND() | Returns an evenly distributed pseudorandom number $>= 0$ and $<1$ |
| ROUND(*n, d*) | Rounds *n* to *d* digits |
| ROW(*ref*) | Returns the row number of a reference |
| SIGN(*n*) | Returns the sign of a number *n* |
| SQRT(*n*) | Returns the square root of a number *n* |
| STDEVP(*n1, n2, . . .*) | Calculates the standard deviation of its arguments |

| | |
|---|---|
| SUM($n1$, $n2$, . . .) | Returns the sum of its arguments |
| SUMSQ($n1$, $n2$, . . .) | Returns the sum of the squares of its arguments |
| TAN($n$) | Returns the tangent of an angle |
| TRANSPOSE(**A**) | Returns the transpose of an array |
| TREND(*known y's, known x's, new x's, constant*) | Returns values along a linear trend by fitting a straight line, using the least squares method |
| VARP($n1$, $n2$, . . .) | Calculates the variance of its arguments |

# Index